# BTEC National for IT Practitioners: Business Units

# BTEC National for IT Practitioners: Business Units

## Core and Specialist Units for the IT and Business Pathway

**Sharon Yull**

AMSTERDAM • BOSTON • HEIDELBERG • LONDON • NEW YORK • OXFORD
PARIS • SAN DIEGO • SAN FRANCISCO • SINGAPORE • SYDNEY • TOKYO

Newnes is an imprint of Elsevier

Newnes is an imprint of Elsevier
Linacre House, Jordan Hill, Oxford OX2 8DP, UK
30 Corporate Drive, Suite 400, Burlington, MA 01803, USA

First edition 2009

**British Library Cataloguing in Publication Data**
A catalogue record for this book is available from the British Library

**Library of Congress Cataloging-in-Publications Data**
A catalog record for this book is available from the Library of Congress

ISBN: 978-0-7506-8662-4

For information on all Newnes publications
visit our web site at books.elsevier.com

Typeset by Macmillan Publishing Solutions
(www.macmillansolutions.com)

Printed and bound in Slovenia
09 10 10 9 8 7 6 5 4 3 2 1

# Dedication

I would like to dedicate this book to my daughter who is a constant source of inspiration.

# Contents

# Preface

## Introduction

Welcome to the ever changing world of business and information and communications technology. This book has been designed to provide you with a range of knowledge, information and skills that will facilitate you in understanding the BTEC Nationals IT Practitioners qualification.

## About the BTEC National Certificate and Diploma

The BTEC National Certificate and Diploma qualifications are Level 3 qualifications that have been designed to provide you with a range of practical skills and underpinning knowledge that will allow you to progress onto a higher level course or prepare you for a job in Business and ICT.

ICT is such a growing area that you will find all areas of the BTEC specification appropriate. The units have been designed with the support of practitioners, experts in the field and also in collaboration with industry. You will be able to use elements of the qualification in a range of situations, whether it is working within a project environment, engaging with e-commerce technology, analysing business systems, undertaking marketing activities, or just having an awareness of the impact that ICT has on organisations.

You do not have to have an extensive knowledge of ICT to embark on the BTEC National qualification. Each of the units provides a good coverage of the subject matter. In conjunction, this accompanying book provides additional support in terms of a range of activities, case studies and test-your-knowledge sections alongside more comprehensive information that follows the guidelines of the specification.

The range of units available on the BTEC Nationals IT Practitioner qualification is quite diverse. The units provide opportunities for you to study at a very specialist level focussing on e-commerce, IT project, marketing, exploring business activity, investigating business resources and looking at the impact of the use of IT on business systems.

On successful completion of this qualification the progression opportunities are quite varied, you could progress onto a Higher National Diploma, Foundation Degree or a Degree programme. Alternatively, you could apply for careers in the areas of business ICT or computing.

## How to use this book

This book provides a support mechanism for the BTEC Nationals IT Practitioner specifications. A range of core and specialist units have been covered within the text, each chapter providing a range of additional materials, activities, case studies and test your knowledge sections.

Each chapter begins with an overview of the content of the related unit and addresses the learning outcomes. Following on from this each of the main headings provides detailed coverage of the learning outcomes.

The activities have been designed to establish your level of learning and provide further opportunities for you to develop your understanding of a specific topic area or concept. The activities are devised to be used at an 'individual', 'group' or 'practical' level. The activities are broken down into a range of tasks that require you to undertake research, develop an understanding, provide an opinion, carry out an activity, discuss and present information.

The question sections provide you with an opportunity to re-visit and refresh your understanding of a previous topic.

In some areas of the book certain terminology is used that you may be unfamiliar with. To support your understanding of this, sections have been included that provide clarification or a definition of the terms referred to.

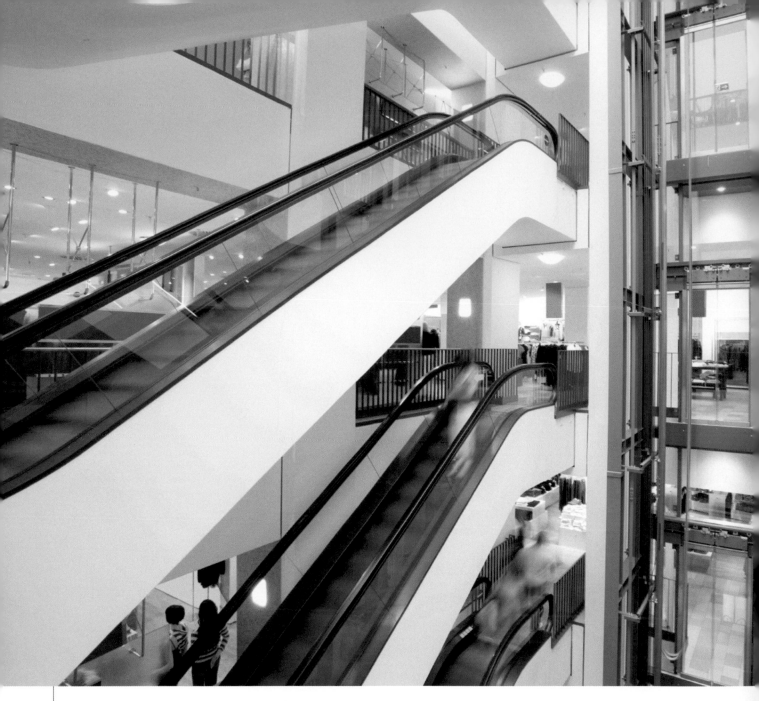

Courtesy of iStockphoto, Bim, Image# 7195255

Organisations engage in a range of business activities on a day-to-day basis. These activities focus on a number components and resources working together to ensure that strategic aims and objectives are being met.

Business activities are also influenced by external factors and third parties such as stakeholders who can contribute to the decision-making process and possibly determine some of the key drivers of the organisation.

# Chapter 1

# Exploring Business Activity

This chapter introduces students to a range of principles and activities, and provides a taster of how organizations are structured, how they function, and the interaction that exists between various functional departments and stakeholders.

Students will gain an insight into what influences a business and external factors that have an impact on a business. It is hoped that students will also consider their own experiences as part-time employees or customers and be able to apply their knowledge to enrich their understanding of business concepts.

This chapter will be structured around the following learning outcomes:

- Understand the different types of business activity and ownership.
- Understand how the type of business influences the setting of strategic aims and objectives.
- Understand functional activities and organizational structure.
- Know how external factors in the business environment impact on organizations.

## Understand the different types of business activity and ownership

Businesses are set up for a range of different purposes and to meet a number of objectives that could be profit, brand, environment or customer driven. All businesses can be categorized in terms of their type: what they are and what they do; and their purpose: why do they exist? Businesses also have common features in terms of ownership and stakeholders, i.e. whom they are accountable to.

### Types of business activity

Organizations can be categorized by the nature of their business and the type of business activity that they undertake. Business activities can be classed as being local, national, international, global, public or private sector, non-profit making and voluntary.

Local and national business activity will be dictated by the nature of the product or service on offer, the marketplace in terms of supply and demand, and the customer and supplier base. A local business could be defined by its geographical boundaries, so for example a fresh fruit and vegetable shop that is based in a small village community may have a geographical boundary of 40 or 50 miles. A national business would have a much wider span of business activity. This could be due to the allocation of resources and logistics, for example Halfords (Figure 1.1).

430 stores nationwide

Figure 1.1  Halfords
http://www.halfords.com

An international business trades across many countries, in terms of importing and/or exporting products or services; however, it does not have any investments outside their own country.

Global businesses such as McDonald's or Coca Cola™ have a presence and investments in a number of countries, with all of their business activities being branded under one name.

---

**Activity 1.1**

Conduct research to provide case study examples of a local, a national, an international and a global organization, and identify their primary business activities.

---

Public sector organizations rotate around government departments (Figure 1.2), each public sector function operating as its own independent organization. Public sector functions rotate around the needs of the public and assist in providing services that provide basic facilities such as healthcare and education. Public sector functions

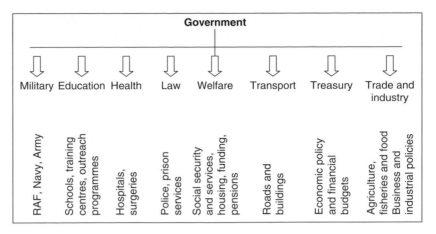

**Figure 1.2** Public sector departments

are also considered as free provisions, although services are paid for indirectly through taxes and contributions.

Other areas that should also be considered under the public sector include environmental agencies, planning and building, and public functions such as emergency services.

The second category of organizations is private sector. The key objective for private sector organizations is to make a profit. These organizations are set up to provide a service or trade in a number of areas, including:

- finance
- insurance
- law
- retail
- manufacturing
- agriculture
- education and training
- technology
- logistics.

Mutual organizations are set up for the benefit of others in that they provide a cooperative function. Examples of mutual benefit organizations are trade unions, clubs and societies. Mutual organizations can be profit making or non-profit making. In some cases the members of the mutual organization will share the benefits and profits that are made.

The final category of organization, which is classified as non-profit making, but does not fall under the umbrella of government public sector or private sector, is that of charities. Charities provide a social service similar to that of government utilities; however, they rely on donations to support their infrastructure, whereas the government relies on public contributions.

Within the above categories, organizations can be classified further by their overall function. These functions can be grouped into one of four areas (Figure 1.3):

- Primary organizations include those of farming and agriculture, e.g. those that are dependent on natural land or sea resources.

CHAPTER 1

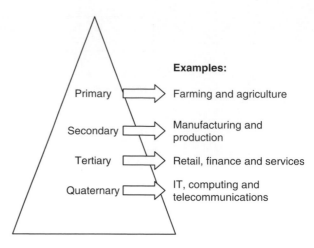

Figure 1.3 Organization groupings

- Secondary industries focus on manufacturing and production industries.
- Tertiary sectors are those that concentrate on retail and providing a service.
- Quaternary organizations have developed and grown over recent years to include sectors such as computing, Internet and web-based companies, and telecommunication businesses.

## Activity 1.2

1. **Provide a real-life example for each of the following sectors of business:**
   - **primary**
   - **secondary**
   - **tertiary.**
2. **Complete the table to identify into which category each of the following organizations would fall:**

| Organization | Category |
| --- | --- |
| An oil refining company | |
| A college | |
| A fast-food retailer | |
| A hospital | |
| A car company | |
| A fisherman | |
| A supermarket | |

## Business purposes

Businesses exist for a variety of reasons; they all have their own distinctive identity, aims, objectives and strategies. Businesses also differ in terms of:

- what they do (product or service based)
- how they do it (selling strategies, prices, meeting supply and demand)
- where they do it (geographically – locally, nationally or globally)
- to whom they sell/who they serve (their customer or client base).

## Case Study 1.1

## Elsevier Publishing

Elsevier is the world's leading publisher of science and health information, with over 2000 journals, 17,000 books, and 1900 new books being produced each year, the subject range extending across physical, health, social and life sciences (Figure 1.4).

Elsevier is a worldwide company, with a global customer base. The company has a scholarly community of 7000 journal editors, 70,000 editorial board members, 200,000 reviewers and 500,000 authors.

## Activity 1.3

**Carry out an investigation into two organizations and provide fact sheets that identify:**

- **what their business purpose is**
- **what product or service they have**
- **who their customer base/target audience(s) is**
- **how they supply their product or service.**

Profit is one of the main drivers behind why a business exists. As a result of this almost every aspect and business resource has to be scrutinized to see whether it is cost-effective and beneficial to the organization in terms of making or retaining money.

The core activity of the business in terms of supplying a product or a service also has to be analysed to see whether or not it is being manufactured and marketed at the right price, level and quantity. At the very minimum an organization has to break even to cover all of the associated overheads. However, to make a profit and invest in the future an organization needs to be supplied at a profitable level.

Some organizations that trade on a local, national or global level may choose to sell certain goods and services either 'at cost' or 'below cost'. The reasons for doing this may include:

- to shift a surplus of stock, possibly with a short expiry date, such as fresh produce

**Choose SubjectArea**

⊞ **Agricultural and Biological Sciences**
⊞ **Arts and Humanities**
⊞ **Astronomy, Astrophysics, Space Science**
⊞ **Built Environment**
⊞ **Business, Management and Accounting**
⊞ **Chemical Engineering**
⊞ **Chemistry**
⊞ **Computer Science**
⊞ **Decision Sciences**
⊞ **Dentistry**
⊞ **Drug Discovery**
⊞ **Earth and Planetary Sciences**
⊞ **Economics and Finance**
⊞ **Energy and Power**
⊞ **Engineering and Technology**
⊞ **Environmental Sciences**
⊞ **Forensics**
⊞ **Health Professions**
⊞ **Immunology**
⊞ **Life Sciences**
⊞ **Materials Science**
⊞ **Mathematics**
⊞ **Medicine**
⊞ **Microbiology and Virology**
⊞ **Neuroscience**
⊞ **Nursing**
⊞ **Organisational Behaviour Psychology**
⊞ **Pharmaceutical Science**
⊞ **Pharmacology**
⊞ **Physics**
⊞ **Psychology**
⊞ **Social and Behavioral Sciences**
⊞ **Toxicology**
⊞ **Veterinary Science and Veterinary Medicine**

**Figure 1.4** Subject range content

- to sell a range of discontinued stock, for example last season's product range, before new stock arrives
- to deliberately undercut a competitor
- to saturate the market with a particular brand and to increase awareness as part of a marketing campaign of a new product
- to reach a new customer base or niche by offering a discounted or promotional price.

## Owners

Business ownership can extend across public, private and voluntary sectors, as identified in 'Types of business activity', above. Types of business ownership can be classified as being single (sole trader) through to partnerships, limited companies (private and public), franchises, government departments and agencies, cooperatives and trusts.

A business with a sole trader is usually owned and controlled by a single person, for example a small corner shop or florist.

Partnerships are usually established between two or more people, where the liability and the profits are distributed accordingly. Partnerships are quite common for accounting and law companies.

Private limited company ownership and control is usually overseen by a board of directors who are usually the main shareholders. Public limited companies also have a board of directors; however, because capital is raised from share issue which anybody can buy, there can be a conflict of interest between the board of directors and the shareholders.

Franchises are set up by an individual or individuals who trade under a corporate brand and supply corporate products or services under that franchise contract. A good example of this is McDonald's (Figure 1.5).

**Figure 1.5** McDonald's franchising
http://www.mcdonalds.co.uk

Government departments and agencies all function independently; however, they are all owned by and accountable to local or national government authorities. Examples include:

- the Ministry of Defence
- the Environment Agency
- the Department for Transport.

There are two types of cooperative: worker cooperatives, where members invest their own capital and share resources and expertise, examples of which can be found in industries such as printing or textiles; and retail cooperatives such as the Co-op.

Trusts can be set up, managed and controlled by a board of trustees. Trusts can also be accountable to various other stakeholders and committees, as would be the case for an education trust.

**Activity 1.4**

1. Identify at least one difference between a sole trader, a limited company, a franchise, a cooperative and a trust.
2. Carry out research to find an example of each of these types of organization ownership.

CHAPTER 1

## Key stakeholders

All businesses are answerable to stakeholders. Stakeholders can be internal or external to the business, and they are people or organizations that have an impact on the way in which a business operates.

Stakeholders can include:

- customers
- employees
- suppliers
- owners
- pressure groups
- trade unions
- employer associations
- local and national communities
- governments
- links and interdependencies.

Depending on their importance to the organization, stakeholders can influence the way in which the business functions and what it sells.

Customers, for example, can put pressure on manufacturers and retailers leading to an increase in more 'organic' or 'green' products and services. Pressure groups can influence policies and procedures, for example the location of a new bypass or housing estate. Employer associations can influence decisions made about the welfare and well-being of staff in terms of common rooms, catering and social activities.

Stakeholders can influence almost any element of an organization, including price, policy, practices, products, services and the environment in which it operates.

---

### Case Study 1.2

### Anglo American

The following case study discusses how Anglo American has adhered to environmental pressures to promote and encourage sustainable development.

### Introduction

Anglo American is one of the twenty largest UK-based companies. It is one of the world's leading international mining companies, employing 120,000 worldwide and 8000 people in Britain. Most of its operations are in the primary sector, for example, extracting materials through mining and quarrying. Anglo American tries to operate in a way that promotes sustainable development through having a positive effect on different types of 'capital':

- social – increasing the life chances of people in local communities
- natural – reducing energy use and the impact of operations on nature
- manufactured – providing infrastructure; coping with waste

- human – supporting education and investing in skills
- financial – concentrating on the whole financial performance, not just returns to shareholders.

## Sustainable development

Sustainable development means not depleting stocks of these types of capital, but passing on at least as much as our generation started with. It shows respect for everyone who shares the planet now or in the future and respect for nature itself. As a major supplier of raw materials, Anglo American tries to balance its use of natural capital (e.g. ore deposits or water used in processing) by increasing human and social capital. To help, it has created a set of principles to meet targets of sustainable development.

## In practice

When Anglo American carries out its mining operations, it tries to have a positive effect on three areas:

1. In the area where the mine is located, it carries out operations with care and tries to improve the life chances of local people.
2. In the area immediately surrounding the mine, it is active in conservation and improvement.
3. In the wider region around the mine, it contributes financially to local communities and helps to generate new businesses and other economic opportunities.

## Pressures

There are pressures to be environmentally responsible from both within and outside the business. Internally, the Board decides how best to work with local communities and mine sensitively. Shareholders are also internal. As the owners of the company they want the business to do well, but to do so, Anglo American must be welcome in the communities where it works or wants to work in the future. By showing that it behaves responsibly, the business is also able to attract the best staff.

Externally, pressures come from legislation designed to protect the environment and from national and international standards and regulations. Customers and other external bodies, like NGOs, also demand high standards of performance.

## Benefits

Acting responsibly gains the trust of communities, respect from governments and the public, and loyalty from customers. It also gives the company an advantage when recruiting new staff. All stakeholders benefit and Anglo American has created a tool (the Socio-Economic Assessment Toolbox or SEAT) to ensure that it consults widely and regularly with local people and measures the impact of its activities in the areas where it operates. It can be used to improve social responsibility and performance.

## Conclusion

As a major mining company, Anglo American recognizes that it has social and environmental responsibilities in the areas in which it operates and that these are fundamental to its business. It has built itself a strong reputation in terms of sustainable operations and through its policies that are carried all the way through the company.

_____

Taken from The Times 100
http://www.thetimes100.co.uk/studies/view-summary-social-environmental-responsibility-65-248.php

## Understand how the type of business influences the setting of strategic aims and objectives

All businesses will have a devised strategy that will provide the framework for what they do and how they do it, whether it is manufacturing a product or selling a service. The strategic planning process may be formulated over months or even years depending on the type and size of the organization.

Strategic planning may differ between private and public sector organizations. The emphasis may be more profit or market-share driven for the private sector and customer service or loyalty driven for a public sector organization. This section will address some of these points.

### Strategic planning process

All organizations have to undertake strategic planning to identify their current position, to reflect on past influences, and to move forward with new strategies for growth, profit, environmental, ethical, market share or customer-focused objectives.

#### Quantitative and qualitative analysis

Planning may involve the collection of quantitative and qualitative data and feedback from users or customers to improve a facility or service. For the strategic planning process to be effective an analysis of the current situation has to be made. This analysis could be based on:

- **Financial status** – how profitable is the company?
- **Environmental issues** – how is the market or specific products and services being affected by the environment: weather conditions or 'greener' issues?
- **Customer opinion** – what do customers think about a certain product or service or the company?
- **Market research** – what are the current trends, what are people buying, what are people prepared to pay?
- **Competitors**.

#### Setting aims and objectives

Setting aims and objectives is crucial to the strategic planning process. Aims and objectives identify what you would like to achieve and how you intend to achieve it, similarly to a goal or target. Within any strategic planning process aims and objectives will provide the framework in terms of how something will be achieved. For example, an organization may plan to make at least 75% of its packaging recyclable within a five-year period. The aim is the statement of purpose: recycling packaging; and the objective is the measurable target: 75% within five years.

### Planning strategies

Planning is crucial in organizations because it provides measurable progress steps for the future. Without planning you would not know where you are going or recognize when you have got there. See Figure 1.6 for an example of strategic business planning by Network Rail.

The Strategic Business Plan outlines in detail how the rail industry aims to meet the growing expectations of passengers and freight users. This includes:

- Continuing to improve the reliability of train services
- By 2014, over 92% of trains will be running on time
- Continuing to invest in the railway infrastructure
- £30bn is needed for the next five years
- Tackling overcrowding and improving train services on busy routes
- £10bn to enhance the network where it is most needed
- Focusing on affordability by setting challenging but realistic efficiency targets
- By 2014, the cost of running the railway will nearly have been halved over ten years.

**Figure 1.6** Network Rail – strategic business plan
Information and logo taken from the Network Rail website: http://www.networkrail.co.uk/aspx/4357.aspx

There is a number of ways to plan in an organization, the way in which you plan being dependent on a number of factors. These factors evolve around a TROPIC cycle (Figure 1.7):

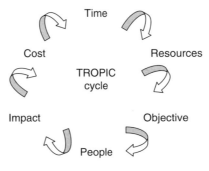

**Figure 1.7** TROPIC cycle of planning

- **Time** – how long is the plan?
- **Resources** – what is going to support the plan?
- **Objective** – what is the purpose of the plan?
- **People** – who is involved with plan?

- **Impact** – who and what will the plan effect?
- **Cost** – how much is it going to cost to implement the plan?

If consideration is not given to all of the factors in the TROPIC cycle a plan could be incomplete and the chances of successful implementation will be greatly reduced. What would be the point of generating a plan to network an entire department if you could not afford the hardware or software, or if you could not implement the network plan within a certain period?

Within an organization planning can be carried out over certain periods, which can be recognized as being short, medium or long term, as identified in Table 1.1.

Table 1.1 Examples of planning

| Type of planning | Period | Example of plan |
| --- | --- | --- |
| Short | 1–3 years | To ensure that all functional departments within the organization are fully networked |
| Medium | 3–5 years | To expand the range of wireless and mobile technologies to all essential users |
| Long | 5 years+ | To implement a range of e-strategies in terms of commerce and procurement |

Strategic planning will allow an organization to conduct both a qualitative and a quantitative analysis of the current situation. Planning will also enable an organization to address a range of aims and objectives, and identify where and how changes can be made to implement these aims and objectives satisfactorily and make the organization more profitable, customer focused or established within the market.

## Public and voluntary sector strategies

The service provision for public sector and voluntary sector organizations differs from that in private sector organizations as the focus will be more customer and service driven, rather than profit or growth driven.

### Public sector and voluntary sector services

Public sector and voluntary sector services cover a wide range of organizations, including primary healthcare trusts, some educational establishments, local government, some transport systems and charities.

The aims and objectives of any strategy for a public or voluntary sector organization will revolve around the customer and meeting their needs or improving the service offered to them. Examples include being able to see more patients within a twenty-four-hour period, or being able to offer a wider selection of BTEC National courses within a college. Voluntary sector organizations may have more global aims and objectives, for example see Figure 1.8.

**Figure 1.8** Oxfam – what they do
http:// www.oxfam.org.uk

---

### Activity 1.5

Oxfam is a primary example of a global voluntary sector organization, with various strategies aimed at providing support and physical resources to meet a number of needs.

1. Visit the Oxfam website (http://www.oxfam.org.uk) and have a look under the section 'Issues we work on'. Try to identify some of the strategies, aims and objectives for each of the causes that they support.
2. How and in what ways, in your opinion, is Oxfam able to support its infrastructure and meet these aims and objectives?

---

At the heart of public sector and voluntary sector strategies is the need to improve the level of service offered to customers. Depending on the type of organization this could be driven by an improvement to service-level agreements (SLAs) that outline the terms and conditions of the service provision. Quality assurance is also pivotal in terms of ensuring that the service offered meets required standards and legislation. A final consideration could be based around cost, therefore ensuring that any service provided is either at or below cost, so that no profit is made.

## Private sector strategies

Private sector strategies may revolve around profit, increased market share, growth, diversification into new markets or branding. Most organizations have at least to break even to cover their costs. Therefore products or services that they sell must be priced accordingly to ensure that the amount that they sell covers the organization's outlay for producing, marketing and distributing it.

**Service-level agreement (SLA)** – an SLA sets out the terms, conditions and boundaries between different parties. The agreement formalizes the level of service that will be provided, and outlines the role and expectations of the customer and the service provider.

Various strategies may be implemented to maximize profit. These strategies may include looking at their sales, in terms of performance, markets, trends and forecasting for the future. Extending the customer base and reaching out to new target audiences is another strategy that could be implemented to maximize sales and profit. An organization may also look towards its product range to identify whether or not it is still popular with customers and whether it is still selling at the required level to make it profitable.

In terms of survival an organization may rebrand or launch a new promotional or marketing campaign, thus creating greater awareness.

An organization could look to increasing their revenue by reducing its overheads to save money and maximize the profit margin.

Selling enough products just to 'break even' – cover costs of production, marketing and distribution – will not maximize profits. Strategies will need to be employed to ensure that the price of the product and the quantity sold enable an organization to make a profit. Breaking even may maintain the current status of an organization, but without profit growth will be stifled.

Survival in the market is a blend of ingredients that in combination make the perfect recipe for success. These ingredients include:

- pricing and selling strategies
- product strategies
- customer strategies
- finance strategies – revenue, breakeven and costs
- survival strategies – analysis of competitors and the market.

## Growth

Growth of an organization can be measured in terms of profit, sales and market share. How much money has been or is projected to be made, how many products are currently being sold and forecast for the future, and what is the current market share status and how can this be increased?

Within a business growth can be defined and measured in a number of ways. Growth can refer to physical growth and expansion of the business geographically, acquiring new premises, buildings or resources such as people and equipment. Growth can also refer to diversification: movement into another different market, for example a supermarket that traditionally sells food items also selling clothes and homeware items, and in some cases also diversifying into the financial market to offer loans, mortgages and insurance.

Growth can be directed at the market to increase market share and become more competitive. It can be related to the growth of a product or service and the sales depending on supply and demand. At certain times of the year demand for a product or service may grow because of a seasonal influence or other influences, for example an increased demand for certain food items because of celebrity chefs using and promoting them on television (Case study 1.3).

**Case Study 1.3**

THE SCOTSMAN

### Shoppers shell out for chance to join Ramsay live

By Lindsay McIntosh

Supermarkets are bracing themselves for the final rush as budding cordons bleus pull out the stops to join a certain short-tempered celebrity chef in a 'cook-along' that's captured the imagination of the nation.

Hand-caught scallops, sirloin steaks and dark chocolate have been flying off the shelves – all thanks to Michelin-starred master chef Gordon Ramsay.

Fridges and freezers across Britain are stocked up so their owners can cook along with Ramsay tonight as he prepares a three-course meal live on television.

The meal for four kicks off with a starter of pan-roasted scallops with a tomato and herb salsa.

The main course is to be steak and chips with a rocket and parmesan salad, and the indulgent dessert will be chocolate mousse.

Dieters should probably turn off at this point – among the ingredients are dark chocolate, double cream, icing sugar, coffee liqueur and two chocolate-covered honeycombs.

The mass social experiment reflects the messiah-like status celebrity chefs have gained, with thousands gearing up to imitate Ramsay's every action.

18 January 2008
The Scotsman
http://thescotsman.scotsman.com/uk/Shoppers-shell-out-for-chance.3686959.jp

Growth can be temporary depending on supply and demand. It could be in the medium term, for example increased profit growths for a quarter- or half-year return, or it could be long term with plans to expand a business over a number of years.

## Understand functional activities and organizational structure

Organizations can be classified in many ways according to what they do (make a product or provide a service), their motivation (to make a profit or satisfy a particular need), and in terms of how they are structured.

### Organizational structures

All organizations have a particular structure, which may be formally or informally adopted. A formal structure indicates that there are several employee levels within the company which are easily recognized so that individuals know where they are in relation to other employees working within the same department, functional area or branch, and to whom they could report if required.

An informal structure may not necessarily be clearly defined and would usually apply to small organizations with only a handful of employees, each knowing to whom they would report, without the need for an organizational chart or diagram.

Traditionally, there are two types of organizational structures, referred to as:

- tall/hierarchical
- flat.

Organizations that adopt a tall structure tend to be quite large in size. The tall structure refers to the number of layers within the organization, making it tall/hierarchical in origin, each layer within the structure representing a level of management or user (Figure 1.9).

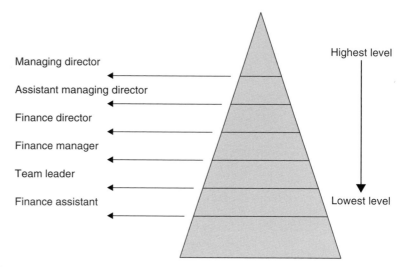

Figure 1.9 Tall structure

Organizations can also be classified as flat structured (Figure 1.10). Flat-structured organizations tend to be smaller, with a smaller span of control and fewer management levels.

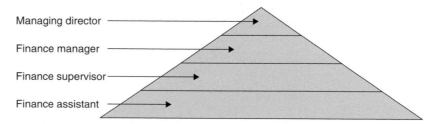

Figure 1.10 Flat structure organization

Over recent years some organizations have chosen to adopt a more flat-style structure because of the benefits that it can bring. This transition to a less rigorous structure is referred to as 'delayering'.

Delayering will remove some of the more senior levels of management (Figure 1.11), thus improving the lines of communication between the

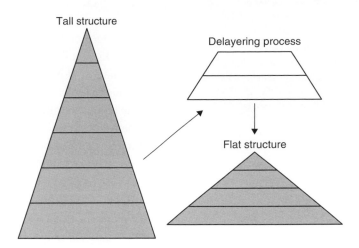

**Figure 1.11** Delayering process

lower and upper levels of users and management. In addition, delayering can introduce a greater level of flexibility within the organization and remove some of the more formal and rigid boundaries. This greater corporate flexibility can facilitate new, more dynamic opportunities in terms of functional activities and job roles. Delayering can also impact on profits and costs, reducing certain overheads by streamlining procedures.

Inevitably, an organization may find that a restructuring of the organization could make them more competitive at all levels, and provide them with an advantage over organizations that offer the same products or services.

Flat- and tall-structure organizations both have benefits and drawbacks. The size of an organization can dictate these structures. It would be impossible for a large multinational company to adopt a very streamlined flat structure because of the need for formality and rigour throughout the management levels. Smaller organizations can be flatter because the span of control is less rigorous.

- Benefits of a tall-structure organization:
  - defined layers make it easier to see who is responsible for what or whom
  - defined functional areas, e.g. sales, finance, human resources
  - defined management structure levels make decision making more localized and applicable to each level.
- Drawbacks of a tall-structure organization:
  - communication between the layers can be time consuming and information can be inaccurate or misleading ('Chinese whispers': by the time the message has passed to other people in the organization it might have changed)
  - decision making could be delayed because approval may be required from people higher up the hierarchy
  - tendency for information to filter down from the top layers (top–down): manager led, not employee led (bottom–up).

> **Span of control** – the control that an individual, e.g. a manager, has over a team of employees. The larger the span of control, the greater the number of employees to manage.
>
> **Delayering** – the process of reducing the number of layers within a tall structure organization to make it flatter. The layers that are usually reduced are those of middle management.

- Benefits of a flat-structure organization:
  - more direct communication between lower and upper levels and vice versa
  - reduced barriers to communication
  - more open communication between the levels.
- Drawbacks of a flat-structure organization:
  - decisions and policies dictated by one or two people, e.g. shop manager/owner
  - role definition may not be clearly defined, e.g. sales administrator, finance clerk
  - functional departments not clearly defined; a small number of people may assume a number of functional roles, e.g. accounts, IT, sales.

---

### Activity 1.6

1. Identify two basic fundamentals of an organizational structure.
2. What are the two types of organizational structure?
3. Give one advantage and one disadvantage of each.
4. What is the process of reducing the structure of an organization?

---

### Organizational charts

An organizational chart usually depicts how employees in an organization, functional area or department are set out, and their relationship with other employees.

The basic fundamentals of an organizational chart or structure (Figure 1.12) include:

- layers of employees extending from the most senior at the top to the least senior at the bottom of the structure

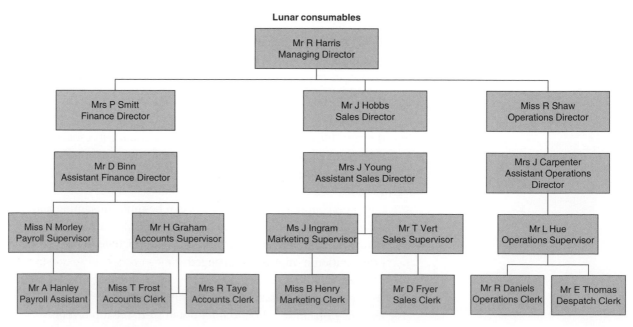

**Figure 1.12** Organizational chart

- names, titles and department/function references for each employee/group of employees
- links depicting how each employee is related to other employees above, below and laterally to them on the structure (chain of command).

A number of observations can be made from Figure 1.12:

- There are three functional departments: finance, sales and operations.
- There are five defined layers of employees, extending from the managing director through to clerks and an assistant.
- An indication is given of employees above, below and laterally in terms of the chain of command within the organization.
- There are more employees at the bottom of the structure than at the top.

---

### Activity 1.7

Based on the organizational chart shown in Figure 1.12, complete the following tasks.

1. Which two supervisors work together as co-workers and which two supervisors work autonomously?
2. Why are there more employees at the lower end of the structure than at the top?
3. Who would the following people need to go through to get to the directors of their departments:
   - Miss T Frost
   - Mr E Thomas
   - Ms J Ingram
   - Mr A Hanley?
4. Why do you think that an organizational structure is needed, especially in large organizations?

---

Organizational structures and charts are important because they provide a formal visual structure of an organization. This is important and beneficial because it becomes transparent in terms of responsibilities, lines of management and communication. If an organization is split across a number of branches or geographical regions it is even more important to have a formal structure as the division of work, roles and responsibilities will be less consolidated.

If an organization has multiple branches or offices, not only will the division of labour and work be split, but functional areas may be divided. For example, all of the key functional departments such as IT, finance and human resources may be located in a head office.

## Functional activities

Organizations are built on a collection of functions, the infrastructure of any organization being dependent on the support provided by functional departments. Functional departments serve a specific purpose within the organization and together the departments enable the organization to achieve its objectives. The most common functional areas of an organization are shown in Figure 1.13.

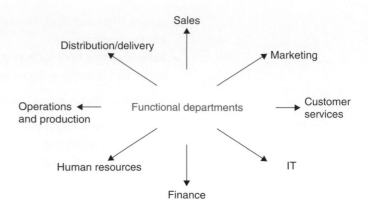

**Figure 1.13** Functional areas

### Sales

The sales department can serve a number of functions, and in some cases provide supporting functions for other departments if there are no dedicated marketing or customer service departments. Sales deal primarily with customers, their aim being to generate orders which may extend from customer enquiries.

### Marketing

The marketing department is sometimes integrated with sales, its role being to advertise and promote the products or services of the organization. This department is involved in the following activities:

- designing and developing promotional materials
- branding and defining the corporate image
- preparing marketing events such as launches and campaigns
- liaising with graphic design or public relations companies if advertising is not developed in-house
- website developments
- scrutinizing competitors' developments.

### Customer services

The customer services department focuses on serving the needs of the customer, acting as a focus for customer enquiries, complaints and appraisals. This department is pivotal to the organization because the feedback received is channelled through to other departments so that they in turn can act on and improve their provisions for customers.

### IT

The IT department provides the technical hardware and software support for the organization. This department can provide a variety of services to an organization, including:

- maintaining the hardware and software systems
- procurement of hardware and software
- managing the telecommunications within the organization
- upgrading systems

- user support and user training
- setting up and maintaining a website
- installations of new hardware and software
- data backups
- integrating and networking functional systems.

An IT department can be further divided into smaller functions, each focusing on a specific area of IT (Figure 1.14).

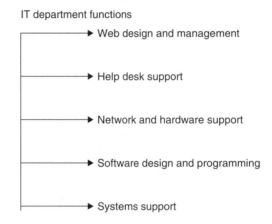

IT department functions

→ Web design and management

→ Help desk support

→ Network and hardware support

→ Software design and programming

→ Systems support

**Figure 1.14** Functional areas of an IT department

### Finance/accounts

This department serves a dual purpose within the organization. The primary function of the finance department is to ensure that there is financial stability within an organization and a steady cash flow to support day-to-day transactions. The finance department is also responsible for:

- payments and transactions
- investments
- accounting procedures (profit and loss, balance sheets, end-of-year accounts)
- payroll
- budgets and forecasting.

### Human resources

Human resources (personnel) provide support to the employees of an organization. The primary function is the welfare of the staff, to ensure that they are advised, guided and motivated to enable them to work productively. This department is also responsible for the hiring and firing of staff. Personnel is usually involved with the restructuring of departments and involved in any delayering activities that may take place.

### Operations and production

Depending on the organization type, one of the functional departments could be operations or production, where all of the physical manufacturing and processing of products or services take place. An operations/production department would certainly be present in large national or multinational manufacturing organizations such as food-processing or car-manufacturing plants.

### Distribution/service delivery

The distribution department or function is concerned with the delivery and logistics of physical stock items. Distribution may not be a discrete department in its own right; it could be linked to operations or sales functions.

Depending on its size, an organization may have functional departments or integrated departments. Smaller organizations may carry out the tasks and responsibilities of a functional department but without the dedicated resources to support it. A business such as a small kiosk selling refreshments would indeed need to sell, promote, obtain stock, distribute the stock and finance it, but all the roles may be delegated to a single person.

Functional departments within an organization need to communicate with each other to ensure that information and good practice are shared throughout the organization. To ensure that this happens effectively and positively certain measures need to be taken. The first is to ensure that a good organizational structure is in place with appropriate levels, as mentioned previously. Secondly, steps need to be taken to ensure that the flow of information between the levels of an organization is open and fluent.

### Other functional areas and activities

Depending on the type of organization, other functional activities may exist, including the need for:

- research and development
- procurement
- logistics
- customer services.

### Management information systems

Management information systems (MIS) can help to improve communication within an organization and support functional departments in their roles.

## Relationships between functional activities

There are several factors and influences to consider when examining the relationship between functional activities.

Within an organization there may be a range of functional activities that are attributed to a specific functional department or departments. Monitoring the workflow of these departments and ensuring that they have adequate resources to support them with their functional activities will enable them to be more efficient and productive.

Some functional activities are interdependent on others, therefore enhancing communication and sharing resources across these activities will ensure that various tasks are completed on time, within budget and to agreed service levels.

Some organizations prefer to outsource some of their functional activities, for example the use of external call centres, outsourcing an IT provision, or using an external marketing organization to promote a product or service. The trend towards outsourcing has grown over the years as this is deemed to be more cost effective and less labour and resource intensive.

---

### Case Study 1.4

### Outsourcing – How Aviva bet big on India

### Positive news from Norwich Union's parent

When Norwich Union announced its offshoring plans at the end of 2003 there was an outcry. The mainstream media notched it up as another sell out by a historic British brand and there were fears customer service would deteriorate.

But, in a rare speech about the whole shift, the man charged with ramping up operations in India has detailed the success so far.

The fact that the decision was made as recently as 2003 now seems bizarre – it feels like a strategy that has been in place for longer. But Sean Egan, Aviva CEO of Offshore Services, yesterday told a conference that when the company had hit a wall in being able to build a call centre and recruit the necessary staff, the die was cast and since that time there have been benefits in terms of improved costs, quality and productivity.

> 'It's like lifting up the carpet and seeing what's underneath.' – Sean Egan, Aviva CEO of Offshore Services, on the hidden benefits of offshoring

Egan said: 'Our greatest successes have not come in back office work but have come – after a bit of pain – in the front office.'

He portrayed the whole move – as one might expect from the UK's largest insurer – as being about risk mitigation. As such, five sites have been used across India, spread out from the north of the country to Bangalore and Chennai in the south and even a facility in neighbouring Sri Lanka.

There are three providers for business process outsourcing (BPO), 247, EXL and WNS, once upon a time BA's captive operation in India which it then spun out.

The two companies used for IT are well-known Indian giants, namely TCS and Wipro.

After initial speculation about job numbers to go to India, Egan revealed that by this year there are around 4,500 employed in BPO and 1,100 in IT. As if to answer some of the critics, he said most of those who lost their positions in the UK were able to stay within the group.

The very precise forecast is that by 2008 there will be 7,607 employed in BPO in India and 1381 in IT outsourcing.

The company opted for a build–operate–transfer approach to offshoring – as opposed to plain offshore outsourcing, building from scratch or a managed services approach. Egan said this has 'allowed the company to ramp up fast' – although he later said that in hindsight expansion should have been more cautious.

He insists that cost, while a 'huge driver', wasn't the only reason for offshoring – though he said savings were an impressive 40 per cent and 'sometimes better than that'. He said Aviva's brands must be competitive in a UK market driven by service levels and that quality is key. There is a well-developed to-ing and fro-ing of staff from India to the UK and vice versa, and he described the buildings and working conditions in India as the best in the whole group.

So where does this leave the UK workforce? According to Egan, they now work in a more 'data rational and analytical space' than they were previously. It's what advocates have long promised about moving up the value chain.

By Tony Hallett, Silicon.com
18 October 2005
http://www.silicon.com/financialservices/0,3800010322,39153441,00.htm

## Activity 1.8

Aviva is one of many organizations that have decided to outsource parts of their provision.

1. What influenced Aviva's decision to outsource?
2. Provide examples of three other organizations that have an outsourcing provision.
3. Why do organizations outsource? What are the benefits of this?

## Influencing factors

Business structures and the functional departments that exist within them can all be influenced by a number of factors such as the size of a business, the environment in which the business operates and the strategic plan of the business.

Large organizations tend to have a rigid framework based on tall structures. Therefore, it could be argued that the business structure is formal because of the number of layers within that business structure, and because the functional departments are very defined in terms of their roles and responsibilities in supporting that structure.

Smaller organizations may not have a formally defined structure and the idea of having functional departments may revolve around one or two people who have responsibility for management, sales, marketing, finance, IT and distribution.

The business environment can influence the organizational structure and functional activities. For example, organizations that operate in dynamic environments, such as retail and supermarkets, where products are bought on an almost continuous basis throughout the day, require support mechanisms to be in place. This ensures that there is adequate stock on the shelves and in the warehouse, that customer demands are being met, that the IT infrastructure can manage the large volumes of data, that promotions are set up and targeted correctly at the right audience, and that prices are competitive to draw in the customer base.

Strategic planning can impact on organizational structures and functional activities, especially if the strategic planning includes the updating, removal or expansion of resources.

## Know how external factors in the business environment impact on organizations

Organizations can be influenced by a range of external factors in the business environment that can have both a positive and a negative impact on every aspect of their business. Some of these factors can be political, economic, social or technological.

### Political factors

Depending on the type of business, political factors may or may not have much of an impact on the arena in which a business operates. Political influences can include:

- national and international law
- government policies.

National and international law can determine and influence which markets you trade with and outline restrictions regarding the import and export of goods and services. In addition, these laws will influence how much you pay your employees – for example minimum wage – and also ensure that employers do not discriminate against employees in terms of age, gender or religion, etc.

National and international law is also there to protect consumers and safeguard them in terms of what they have purchased and their rights as a consumer to return goods and services if faulty or unwanted.

Consumer rights have to be taken into consideration in terms of how this can impact on the environment of a business. Ensuring that products and services meet certain standards in terms of health and safety and being fit for purpose are core considerations.

Environmental issues are quite high on governments' agenda and the drive towards a 'greener' environment has forced many organizations into changing their practices to meet certain government targets (see Case Study 1.5).

---

### Case Study 1.5

### Environmental influences

### Airlines sport their green colours

Every day at Britain's airports, hundreds of aircraft take off for destinations across Europe. Each churns out tonnes of carbon dioxide, a by-product of the jet engine and a likely cause of global warming.

Take just one flight. Ryanair's 800-mile (1,300 km) flight from Stansted to Rome, using a Boeing 737, will produce 27 tonnes of $CO_2$ as it goes.

Much of it will hang around in the atmosphere, contributing to the greenhouse effect.

A flight across Europe can produce 27 tonnes of $CO_2$

The greenhouse gases generated by air travel are tiny compared with many other environmentally damaging human activities.

The Intergovernmental Panel on Climate Change estimates aviation contributes just 3% to total global emissions of $CO_2$, compared with the 25% pumped out by power stations.

But there are predictions that this will rise to 15% because aviation is one of the few sources of greenhouse gases that are growing.

Air travel has been predicted by the government to triple in the next 30 years.

Airports are being expanded to cope with the extra demand, with extra runways planned at Stansted and Heathrow.

Now, enter the combined weight of the UK's aviation industry – with a strategy designed to show that airlines, airports and aircraft manufacturers are taking responsibility for what they admit are the 'significant, detrimental environmental impacts' of our love of flying.

## Economic contribution

The target is to make planes 50% more fuel efficient by 2020, compared with aircraft in our skies now.

That should reduce $CO_2$ emissions by half. But can it be done – and what impact will it have on global warming?

For evidence that it is possible, the industry points out that modern aircraft are 70% more fuel-efficient than they were in the '60s.

Planes can be built much bigger. Airbus says its giant new A380 burns 13% less fuel than the ageing Boeing 747.

A380s are likely to be common at airports in 15 years' time.

Making the air traffic control system more efficient may help, too. Time wasted, on the ground or in the air, is paid for in aviation fuel.

But environmentalists doubt that building better planes with better engines can achieve the 50% target.

Jeff Gazzard, from the Aviation Environment Federation, said the strategy was 'hopelessly optimistic, and over-reliant on technology. Real back-of-the-fag-packet stuff'.

He believes a 25% reduction is possible, but says that will not be enough. If the government's estimates are to be believed, in the 15-year timescale of this strategy the number of flights will increase by 150%.

The aviation industry is also committed to a system of 'emissions trading' which would allow airlines to buy the right to produce greenhouse gases from other industries that are producing less – such as power stations.

Do we need to put the brakes on cheap flights?

This 'virtual pollution market', it is argued, would put a price on environmental damage, and encourage greener air travel.

But environmentalists believe none of these solutions will tackle the real problem: our growing desire to get on a plane and fly, whether on a business trip across the globe, or a cheap trip to a hot new holiday destination in Europe.

The only solution, they say, is to make flying more expensive, to persuade us to fly less.

The Aviation Environment Federation wants every passenger to pay at least £34 more for every 700 miles (1,100 km) they fly.

The aviation industry hits back with figures showing air travel contributes £14 bn a year to the British economy. And the debate continues.

One thing everyone's agreed on is that our love of the high life has an environmental cost, and it is a big problem that needs to be solved.

_____

By Tom Symonds, BBC Transport Correspondent
20 June 2005
BBC News http://news.bbc.co.uk/1/hi/uk/4111310.stm

_____

In addition, new car tax bands have been introduced for vehicles registered after 1 March 2001 to encourage low $CO_2$ emissions. This may impact on supply and demand for certain vehicles supplied by car manufacturers. See Figure 1.15 for the price banding and Figure 1.16 for the 'Fuel economy label'.

Political factors can include changes in government policy, for example white papers on education setting out agendas for qualifications in the future: '14–19 diplomas' and 'vocational learning'; decisions in regard to increasing the school leaving age and policies outlining 'inclusion'.

The government can also provide tax breaks, subsidies and grants to support new businesses, or a business that operates in a specific sector, for example agriculture.

## Economic factors

Economic factors can have a huge impact on organizations. Businesses are influenced by a range of economic factors, including pay levels,

Petrol car (TC48) and diesel car (TC49)

| Band | $CO_2$ emission (g/km) | 12 months rate | 6 months rate |
|------|------------------------|----------------|---------------|
| A | Up to 100 | Not applicable | Not applicable |
| B | 101–120 | £35.00 | Not applicable |
| C | 121–150 | £115.00 | £63.25 |
| D | 151–165 | £140.00 | £77.00 |
| E | 166–185 | £165.00 | £90.75 |
| F | Over 185 | £205.00 | £112.75 |
| G | Over 225 – for cars registered on or after 23/03/06 | £300.00 | £165.00 |

Alternative fuel car (TC59)

| Band | $CO_2$ emission (g/km) | 12 months rate | 6 months rate |
|------|------------------------|----------------|---------------|
| A | Up to 100 | Not applicable | Not applicable |
| B | 101–120 | £15.00 | Not applicable |
| C | 121–150 | £95.00 | £52.25 |
| D | 151–165 | £120.00 | £66.00 |
| E | 166–185 | £145.00 | £79.75 |
| F | Over 185 | £190.00 | £104.50 |
| G | Over 225 – for cars registered on or after 23/03/06 | £285.00 | £156.75 |

Figure 1.15  Car tax price banding
http://www.direct.gov.uk/en/Motoring/OwningAVehicle/HowToTaxYourVehicle/DG_10012524

cost of credit, competitive pressures, globalization markets, labour, supply and demand, and energy prices. Some of these factors can be managed by an organization, and they can budget for price rises and falls, and costs associated with pay levels and labour prices. Some economic factors, however, are beyond the control of an organization and are dependent on world markets and economies, share prices, and supply and demand as dictated by consumers, competitors and suppliers.

## Social factors

Social factors can also have an impact on the business environment of organizations. An ageing population can influence and dictate the demand for certain products. If, for example, a business is set up in an area where there is a large retired community the products or services may need to be orientated towards the needs of this target audience.

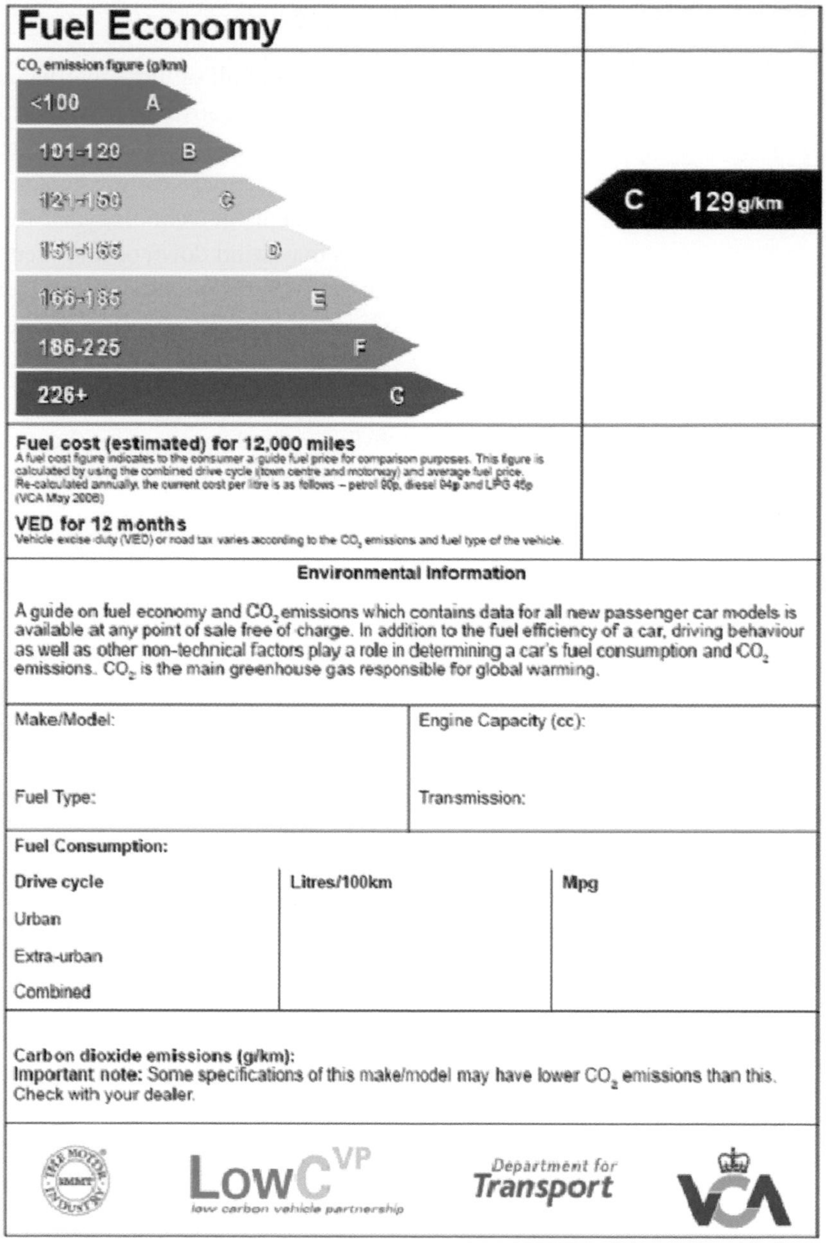

Figure 1.16  Fuel economy label
http://www.dft.gov.uk/ActOnCO2/index.php?q = fuel_economy_sticker

A business can also be influenced by social events, sporting activities, and cultural or celebrity events. For example, the Olympic Games generate a wealth of new business opportunities for local communities and also through merchandising.

## Technological factors

Developments in network communications, broadband and telephony services have had a tremendous impact on businesses. In addition, the Internet and e-mail communications have opened up the market and enabled businesses to reach more global and targeted audiences twenty-four hours a day.

## Impact

The impact of all of these internal and external influences can be both positive and negative. More opportunities now exist for new businesses with easier access to information, resources and customers. Finding new supply sources and raw materials for products is made easier by enhancements in technology. However, as one new business emerges others may wind down or find themselves in a merger or takeover situation driven by costs and efficiency.

Strategic planning and the functional activities of an organization can also be influenced by a range of political, economic, social and technological factors. As a result, organizations must embrace these challenges and demonstrate that they are robust and dynamic in their field to ensure continual growth and longevity in the future.

## Questions and review

1.  Provide an example of a local, a national, an international and a global business.
2.  What are the main sectors of business?
3.  Why, in your opinion, do businesses exist?
4.  For what reasons might a business offer a product or service either at or below cost?
5.  What types of business ownership are there? Provide a brief description of each.
6.  Identify four stakeholders in the following organization types:
    *   hospital
    *   school
    *   travel agency
    *   waste disposal company.
7.  Why is it important for a business to set aims and objectives?
8.  How and why might the strategies of a public sector organization vary from those of a private sector organization?
9.  Provide two examples of how a business can 'grow'.
10. Identify four pieces of information that you could obtain from an organizational chart.
11. What is meant by the term 'span of control'?
12. Identify and describe the key components of five functional areas within an organization.
13. What, in your opinion, are the benefits of outsourcing?
14. How can political factors impact on the business environment of organizations?
15. How can supply and demand impact on the business environment of organizations?
16. Social influences can impact on the business environment of organizations in a number of ways, for example hosting of a major sporting or cultural event. Explain how, for example, the Olympics can impact on the business environment of an organization(s).
17. Why and in what ways would advances in technology impact on a business environment?
18. Takeovers and mergers are one way in which an organization can combine resources, reduce costs and demonstrate growth within a market. What impact can a takeover or merger have on a business organization and its employees?

## Assessment activities

| Grading criteria | Content | Suggested activity |
|---|---|---|
| **Pass** | | |
| P1 | Describe the type of business, purpose and ownership of two contrasting organizations. | Carry out research on two different organizations, for example a college and a large insurance company (or you could use a place of work). Prepare a report that will cover a range of criteria. The first part of the report should describe the business type, purpose and ownership of both. |
| P2 | Describe the different stakeholders who influence the purpose of two contrasting organizations. | Produce a short presentation describing the different stakeholders that influence the purpose of the two contrasting organizations. |
| P3 | Outline the rationale of the strategic aims and objectives of two contrasting organizations. | The rationale of the strategic aims and objectives for these two organizations could be included within another section of the report. |
| P4 | Describe the functional activities, and their interdependencies in two contrasting organizations. | A further section of the report could describe the functional activities and their interdependencies within the two organizations chosen. |
| P5 | Describe how three external factors are impacting upon the business activities of the selected organizations and their stakeholders. | Produce an introductory seminar sheet describing, with examples, how three external factors impact upon the business activities of selected organizations and their stakeholders. |
| **Merit** | | |
| M1 | Explain the points of view from different stakeholders seeking to influence the strategic aims and objectives of two contrasting organizations. | In conjunction with P2 you could further develop this area of stakeholder views by producing a presentation that explains the points of view from different stakeholders seeking to influence the strategic aims and objectives of either the two organizations selected for the report, or for two newly researched organizations. |
| M2 | Compare the factors which influence the development of the internal structures and functional activities of two contrasting organization. | In conjunction with P1, P3 and P4 the final section in the report could compare the factors that influence the development of internal structures and functional activities of two contrasting organizations. |
| M3 | Analyse how external factors have impacted on the two contrasting organizations. | Prepare a seminar session aimed at SME's that analyses how external factors have impacted upon two contrasting organizations. Present the information in either a written information leaflet/case study format or in a presentation format. |
| **Distinction** | | |
| D1 | Evaluate how external factors, over a specified future period, may impact on the business activities, strategy, internal structures, functional activities and stakeholders of a specified organization. | In conjunction with M3, develop the seminar session further to include additional information that evaluates how external factors, over a specified future period, could impact upon business activities, strategy, internal structures, functional activities and stakeholders of a specified organisation. |

Courtesy of iStockphoto, fotoVoyager, Image# 1385362

The resources that are required to support a business can be identified as human, physical, technological and financial. Each of these resources impact upon a business in different ways, and it is the management of these resources that can determine the success or failure of the business.

Human resources include having the skills and expertise to address a range of tasks within a business at all levels. In addition, there has to be sufficient human resources to undertake the tasks required on a day-to-day basis.

Physical resources provide the foundation for a business in terms of buildings, plant, machinery and equipment.

Technological resources and the integration of IT to improve efficiency or productivity or to become more competitive or cost-effective should also be managed effectively as failure to do so could result in operations being disabled within a business at the cost of hundreds, thousands or even millions of pounds.

Financial resources and an awareness of finance sources, budgets, profitability, solvency and other accounting measures is also required to secure the current and future prospects of a business.

# Chapter 2

# Investigating Business Resources

There are four resources that an organization needs to manage successfully in order to function effectively. These resources can be categorized under human, physical, technological and financial.

This chapter will examine how each of these resources impacts on organizations and how they support day-to-day operations.

In terms of the human resources, we will explore how people, employers and teams contribute to the growth and wellbeing of an organization. Physical resources will look at tangible resources such as the actual building of the organization and any plant, machinery or equipment that is used. Technological resources examine the importance of an organization and how it needs to conform to certain legal requirements and legislation to protect and be protected against data and information theft, i.e. intellectual property rights. Finally, in terms of the financial resource we will examine the importance of finances in terms of what is required for startup, running costs, the importance of budgets, and the need for and use of other financial reporting documents such as balance sheets and profit and loss accounts.

This chapter will be based around the following learning outcomes:

- Know how human resources are managed.
- Understand the purpose of managing physical and technological resources.
- Understand how to access sources of finance.
- Be able to interpret financial statements.

## Know how human resources are managed

Human resources are pivotal to organizations in terms of the people, employers, teams and stakeholders who have a vested interest in the growth and success of the organization.

### Human resources

Staff and employees are required to ensure that a business functions and communicates effectively with a multitude of resources. Staff are required to meet the basic business requirements, needs and demands as dictated by customers, employers and other stakeholders.

Within an organization there are employees at different levels, working in different functional departments and having different roles and responsibilities. Each employee may work individually in the achievement of various tasks; however, a large majority of organizations work on a team basis where different people in different departments work together to share skills and knowledge on a common project.

Teams can be a very effective way of achieving tasks. The combination of skills and expertise can ensure that targets are met and projects signed off.

### *Characteristics of teams*

A number of factors can contribute to the effectiveness of a team:

- **The purpose of the team** – what have they been brought together for?
- **The environment in which the team has to operate** – is it a formal work environment or an informal social environment?
- **The constraints imposed on the team** – timescales, budgets, resources, etc.
- **The characteristics of the team** – who are the individuals, what skills and traits do they bring to the group, etc.?

The characteristics of team members are a very influential factor in how successfully or productively a group bonds. Belbin (1981) suggests that an effective team is made up of people who fulfil a number of roles, as shown in Table 2.1.

Larson and LaFasto (1989) identified effective teamwork attributes, by identifying eight characteristics that could help to ensure the success of a group project:

- **a clear, evaluating goal** – a sense of mission being created through the development of an objective which is understood, important, worthwhile and challenging
- **a results-driven structure** – the structure and composition of the team should be commensurate with the task being undertaken
- **competent team members** – a balance of personal and technical competence
- **unified commitment** – creating an environment of 'doing what has to be done to succeed'
- **fostering a collaborative climate** – encouraging reliance on others within the team
- **standards of excellence** – through individual standards, team pressure and knowledge of the consequences of failure

Table 2.1  Belbin's team roles

| Roles and Descriptions | | |
| --- | --- | --- |
| Team-Role | Contribution | Allowable Weaknesses |
| Plant | Creative, imaginative, unorthodox.Solves difficult problems. | Ignores incidentals. Too pre-occupied to communicate effectively. |
| Resource Investigator | Extrovert, enthusiastic, communicative. Explores opportunities. Develops contacts. | Over-optimistic. Loses interest once initial enthusiasm has passed. |
| Co-ordinator | Mature, confident, a good chairperson. Clarifies goals, promotes decision-making, delegates well. | Can be seen as manipulative.Offloads personal work. |
| Shaper | Challenging, dynamic, thrives on pressure. The drive and courage to overcome obstacles. | Prone to provocation. Offends people's feelings. |
| Monitor Evaluator | Sober, strategic and discerning. Sees all options. Judges accurately. | Lacks drive and ability to inspire others. |
| Teamworker | Co-operative, mild, perceptive and diplomatic. Listens, builds, averts friction. | Indecisive in crunch situations. |
| Implementer | Disciplined, reliable, conservative and efficient. Turns ideas into practical actions. | Somewhat inflexible. Slow to respond to new possibilities. |
| Completer Finisher | Painstaking, conscientious, anxious. Searches out errors and omissions. Polishes and perfects. | Inclined to worry unduly. Reluctant to delegate. |
| Specialist | Single-minded, self-starting, dedicated. Provides knowledge and skills in rare supply. | Contributes on only a narrow front. Dwells on technicalities. |

- **external support and recognition** – where good work is performed, recognize it
- **instituting principled leadership**.

To get the most from team working the following points should be considered:

- **T** – Talk and communicate with other group members continuously.
- **E** – Enjoy the experience and relax into any given role.
- **A** – Ask questions to ensure that you understand what is expected of you.
- **M** – Motivate yourself and others.
- **W** – Welcome thoughts and ideas from other team members.

- **O** – Offer your skills and knowledge willingly for the benefit of the group.
- **R** – Review and reflect on what you have done and how you could have improved on your contribution.
- **K** – Keep within the framework of any timescales and deadlines.
- **I** – Include others in any decision-making process.
- **N** – Notify the team of any problems sooner rather than later.
- **G** – Give 100% at all times.

With any employee working as an individual on sole tasks or within a group, it is important to ensure that there is an awareness and appreciation of other business functions and how these all come together to ensure the success and growth of the organization. Communication and liaison with other functional departments is essential to ensure that developments and good practice are shared across the organization.

## Activity 2.1

1. **Do you think that team working is an important element within organizations?**
2. **Why do you say this?**
3. **Can you provide any examples of where you have contributed to a task that was team based?**
4. **What was the outcome of this task? Was it a success?**

## Developing a professional culture

As an employee you have a professional responsibility to behave and to represent your organization in a professional way. Some employees have to adhere to a code of conduct or codes for professional practice to ensure that their conduct at work and the way that they present themselves reflect positively on the organization that employs them. For some professions an employee may also be expected to belong to a professional body and adhere to the rules and regulations as set out by their code of conduct. An example, the BCS Code of Conduct, is presented in Case study 2.1.

### Case study 2.1

### BCS Code of Conduct

British Computer Society
Code of Conduct

Reproduced with permission of The British Computer Society www.bcs.org/

**The Public Interest**..

1. You shall carry out work or study with due care and diligence in accordance with the relevant authority's requirements, and the interests of system users. If your professional judgement is overruled, you shall indicate the likely risks and consequences.

- The crux of the issue here, familiar to all professionals in whatever field, is the potential conflict between full and committed compliance with the relevant authority's wishes, and the independent and considered exercise of your judgement.
- If your judgement is overruled, you are encouraged to seek advice and guidance from a peer or colleague on how best to respond.

2. In your professional role you shall have regard for the public health, safety and environment.
   - This is a general responsibility, which may be governed by legislation, convention or protocol.
   - If in doubt over the appropriate course of action to take in particular circumstances you should seek the counsel of a peer or colleague.

3. You shall have regard to the legitimate rights of third parties.
   - The term 'third party' includes professional colleagues, or possibly competitors, or members of 'the public' who might be affected by an IS project without their being directly aware of its existence.

4. You shall ensure that within your professional field/s you have knowledge and understanding of relevant legislation, regulations and standards, and that you comply with such requirements.
   - As examples, relevant legislation could, in the UK, include The UK Public Disclosure Act, Data Protection or Privacy legislation, Computer Misuse law, legislation concerned with the export or import of technology, possibly for national security reasons, or law relating to intellectual property. This list is not exhaustive, and you should ensure that you are aware of any legislation relevant to your professional responsibilities.
   - In the international context, you should be aware of, and understand, the requirements of law specific to the jurisdiction within which you are working, and, where relevant, to supranational legislation such as EU law and regulation. You should seek specialist advice when necessary.

5. You shall conduct your professional activities without discrimination against clients or colleagues.
   - Grounds of discrimination include race, colour, ethnic origin, sexual orientation.
   - All colleagues have a right to be treated with dignity and respect.
   - You should adhere to relevant law within the jurisdiction where you are working and, if appropriate, the European Convention on Human Rights.
   - You are encouraged to promote equal access to the benefits of IS by all groups in society, and to avoid and reduce 'social exclusion' from IS wherever opportunities arise.

6. You shall reject any offer of bribery or inducement.

## Duty to Relevant Authority

7. You shall avoid any situation that may give rise to a conflict of interest between you and your relevant authority. You shall make full and immediate disclosure to them if any conflict is likely to occur or is seen by a third party as likely to occur.

8. You shall not disclose or authorise to be disclosed, or use for personal gain or to benefit a third party, confidential information except with the permission of your relevant authority, or at the direction of a court of law.

9. You shall not misrepresent or withhold information on the performance of products, systems or services, or take advantage of the lack of relevant knowledge or inexperience of others.

### Duty to the Profession

10. You shall uphold the reputation and good standing of the BCS in particular, and the profession in general, and shall seek to improve professional standards through participation in their development, use and enforcement.
    - As a Member of the BCS you also have a wider responsibility to promote public understanding of IS – its benefits and pitfalls – and, whenever practical, to counter misinformation that brings or could bring the profession into disrepute.
    - You should encourage and support fellow members in their professional development and, where possible, provide opportunities for the professional development of new members, particularly student members. Enlightened mutual assistance between IS professionals furthers the reputation of the profession, and assists individual members.
11. You shall act with integrity in your relationships with all members of the BCS and with members of other professions with whom you work in a professional capacity.
12. You shall have due regard for the possible consequences of your statements on others. You shall not make any public statement in your professional capacity unless you are properly qualified and, where appropriate, authorised to do so. You shall not purport to represent the BCS unless authorised to do so.
    - The offering of an opinion in public, holding oneself out to be an expert in the subject in question, is a major personal responsibility and should not be undertaken lightly.
    - To give an opinion that subsequently proves ill founded is a disservice to the profession, and to the BCS.
13. You shall notify the Society if convicted of a criminal offence or upon becoming bankrupt or disqualified as Company Director.

### Professional Competence and Integrity

14. You shall seek to upgrade your professional knowledge and skill, and shall maintain awareness of technological developments, procedures and standards which are relevant to your field, and encourage your subordinates to do likewise.
15. You shall not claim any level of competence that you do not possess. You shall only offer to do work or provide a service that is within your professional competence.
    - You can self-assess your professional competence for undertaking a particular job or role by asking, for example,
        i. am I familiar with the technology involved, or have I worked with similar technology before?
        ii. have I successfully completed similar assignments or roles in the past?
        iii. can I demonstrate adequate knowledge of the specific business application and requirements successfully to undertake the work?
16. You shall observe the relevant BCS Codes of Practice and all other standards which, in your judgement, are relevant, and you shall encourage your colleagues to do likewise.
17. You shall accept professional responsibility for your work and for the work of colleagues who are defined in a given context as working under your supervision.

## Activity 2.2

1. Do you think that everybody in employment should abide by some sort of professional code of conduct?
2. Can you provide any real-life examples of organizations or professions that employ a code of conduct?
3. Examine one code of conduct and see what the expectations are from employees.

## Formality: professional versus personal expectations at work

Within any working environment there will usually be a certain level of formality where employees will be expected to behave in a certain way. The expectations of an employee at work would be based on professional rather than personal behaviour and policies may be set up to ensure and safeguard against any misuse. This may include the use of the Internet for work-based tasks, as opposed to social surfing (see Case study 2.2).

## Case study 2.2

### University of Manchester

**Use of the Internet**

The primary reason for the provision of Internet access is for the easy retrieval of information for research purposes in order to enhance the ability of its staff to undertake their University role. However, as with email it is legitimate for employees to make use of the Internet in its various forms outside of normal working hours for personal purposes as long as it is not used to view or distribute improper material such as text, messages or images which are derogatory, defamatory or obscene. It is recognised that there can be occasions where it is sensible for the employee to make occasional use of the Internet for personal reasons such as a private transaction, rather than having to spend considerably more time out of the office. Examples of this might include a bank transaction or the booking of a holiday. As long as such personal use is confined to non-working hours, then it is permissible. For employees who do not have defined hours of work, such as academic members of staff, personal Internet use should not interfere, either by its timing or extent, with the performance of the employee's duties.

http://www.campus.manchester.ac.uk/medialibrary/policies/HR/telephone-email-and-internet-use-at-work-policy.pdf

## Incentives and motivation

Some organizations use incentives to motivate employees, and encourage creativity and initiative. Incentives could be in the form of acknowledging and giving praise to an employee or some sort of monetary bonus or a points-based scheme where points can be used towards a range of gifts.

Providing incentives to employees is quite popular in a target-driven environment such as 'sales', where an employer is motivated to exceed any periodic targets by the lure of such incentives.

### Case Study 2.3

### Use of incentives at work

### Incentives That Work Magic!

Managing an office and the multiple personalities of high-strung sales people is a challenge. The most effective managers have found that the easiest incentive ideas work the best to keep these agents on track and your office 'humming.'

### High fives

The best thing that a manager can do for any sales associate is to praise him/her for a job well done in front of his/her peers. That's why walk around management works. You can't wait for the once a week office meeting to acknowledge them, although that is good too, but it has greater impact right at the time that something was accomplished. Agents remember your verbal 'high fives' and repeat them over and over in their minds when they are given at the time when the person is feeling euphoric about his/her success.

The key with these verbal acknowledgments is to say something positive to the agent when other agents are close by so they can hear it as well. Something like, 'did you all hear that Sandy got her first sale today?' Or, 'wow! I heard that you sold the property on Albert Drive that has been on the market for five years. Congratulations!' It doesn't matter what they did, but if it was an accomplishment like saving a sale, solving a big problem, getting a new listing, making a sale, knocking on doors and getting an appointment, cold calling and getting an appointment, let everyone know. Don't overdo, but definitely say something flattering to the agent.

### Small, instant gratifications

A really great incentive is to acknowledge all new pending sales each week in your sales meeting by having your agents grab a silver dollar out of a bank bag filled with silver dollars. If they pick the one that has an 'x' on it, they get an additional $100. It's tangible, fun, instant, and in front of their peers. Once you start this and try to stop, boy will you hear about it from the agents. They love it! Agents save up all their silver dollars at their desk. When they look at them, they are reminded that they are a winner. It has so many lasting effects and such a simple thing to do.

### Award certificates

Every month the office should award Excellence Certificates to agents who have the most listings in units, the most listings in volume, the most sales in units, the most sales in volume, the most closed units, the most closed volume, the top five sales associates in pending volume, and the SOS agent (one who helped another during the month) as a minimum. The more awards and recognition you give, the greater the incentive for the office as a whole. Of course, if you only have a few agents in your office, you need to have fewer awards to give them meaning. However, if you have a large office, 50 or more agents, you should have many awards. And, the reason for the five top sales associates is that you will have more than just one person winning the award each month. Also, these paper certificates look great in their listing presentations. This is a great incentive because you will have all your agents wanting to be part of the team and getting awards.

### Personal notes

A personal note of congratulations or to say 'hang in there' goes a long way. When you write something personal to your agents you will find that it came at just the right time. They will come to you with a simple smile and tell you that they appreciated you taking the time to do it. This is one of those incentives that is quiet, but powerful. When your agents are being recruited by other companies, they will remember all the personal things that you do. It's tougher to get them to make a move.

### Unexpected surprises

When an agent does something really extraordinary, acknowledge them with a gift certificate for dinner and tickets to a special event. They love the attention and it means so much more because it was not expected. A small surprise for an agent who is feeling really down like paying for a mailing for him this month also has great impact. Make these private indulgences. Watch the results!

Giving praise and recognition on a regular basis will have your bottom line singing a sweet tune. Incentives work! Everyone likes to be acknowledged for hard work and it makes one want to work even harder. Give to your agents and they will give back to you ten fold just like magic.

By Patti Brotherton
17 August 2000
http://realtytimes.com/rtpages/20000817_incentives.htm

### Activity 2.3

1.  **Read the article and discuss what sort of incentive schemes are available to employees.**
2.  **Carry out research and identify two other examples where incentives are used to motivate employees.**

## Outsourcing versus in-house decisions

Outsourcing of skills and resources is a move that quite a few organizations have made over recent years in a bid to become more efficient and/or cost effective. A good example of this is the migration of customer service provisions from the UK to countries such as India where the provision is less expensive to manage (see Case Study 2.4).

CHAPTER 2

---

**Case Study** 2.4

## BT call centre provisions

http://news.bbc.co.uk/1/hi/business/2828391.stm 7 March 2003

### BT opens Indian call centres

Telecoms giant BT has joined the growing band of Western firms to transfer call centre operations to India, at a cost of 2,000 UK jobs.

The firm has said it is to open two Indian call centres – in Bangalore and Delhi – employing 2,200 people by 2004.

Call centre work is highly mobile

The shake-up is part of plans which will see BT's UK-based centres drop from about 100 to 31, with the number of workers falling from 16,000 to 14,000.

BT said no permanent UK employees would be made redundant as a result of the move, and there would be no compulsory lay-offs among agency staff.

### Union wrath

But the move has angered UK unions, which have threatened strike action over the switch.

BT was spearheading moves to 'destroy' UK jobs, said the Communication Workers' Union, which dismissed as a 'red herring' the firm's jobs protection pledge.

"In other countries, established providers such as
BT have lost up to 40% of market share.
We will not allow that to happen to us."
Pierre Danon, BT Retail

'The point is that this is UK work, carried out for UK customers, offering a UK service,' said Jeannie Drake, the CWU deputy general secretary.

'Doesn't that sound like work we should be doing in this country?'

According to one report, a series of demonstrations is planned at calls centres nationwide on Friday.

### Investment plans

Pierre Danon, chief executive of BT Retail, said he realised the shake-up was an 'emotional' subject, and said that there was some 'nervousness' over a move which could affect some of BT's 6,000 agency workers.

But he added that the shake-up would see millions of pounds invested in UK call centres.

'It is a tribute to BT that we are maintaining the investment and maintaining employment in the UK,' he said.

'In other countries, such as Ireland and Germany, established providers such as BT have lost up to 40% of market share.

'We will not allow that to happen to us.'

### Elocution lessons

BT is the latest in a series of firms, including insurers Aviva and Prudential, to investigate locating call centres in India, where costs can be 30% lower than in the UK.

The Indian call centre industry employs more than 100,000 workers, who, if serving a foreign market, are often trained in the culture and customs of the country they service.

Workers serving UK customers are often given accent training, and taught about pubs, football and running story lines in popular soap operas, to be able to hold conversations with British customers.

BT's Indian call centres will deal with tasks such as ringing UK customers to remind them to pay their bills.

BT shares stood 2.75p lower at 154p by lunchtime in London on Friday.

Outsourcing is also very apparent in specialist areas such as IT where some organizations rely on third party companies to provide remote IT support and onsite visits in a bid to cut costs in having their own IT department.

## Maintenance of operation

All organizations have to ensure that they have adequate resources to meet their business's task requirements and to ensure that they function at optimum levels of efficiency and productivity. These resources could include sufficient staffing, equipment, working capital, facilities and administrative support. Once in place, these resources also have to be monitored, and trouble shooting or problem solving may be used to control the effectiveness of the resource.

## Understand the purpose of managing physical and technological resources

Physical and technological resources are crucial to organizations, in conjunction with other resources.

### Physical resources

Physical resources are important to organizations at various scales. For example, large manufacturing organizations rely heavily on buildings, factories, plants and warehouses to physically make and store products, for example car plants, food processing companies, clothing manufacturers and engineering companies. Other companies may not require building facilities on this scale, with possibly just a retail outlet to provide a high-street presence.

Some organizations, especially those that rely heavily on e-commerce as their means of trading, may not have any physical buildings.

Materials and waste are resources that will have more importance to manufacturing-type organizations where raw materials are required to make a final product from which consideration would need to be given to wastage, plant and machinery, equipment including ICT, planned maintenance and refurbishment, emergency provision, insurance and security.

## Technological resources: intellectual property

Intellectual property (IP) allows you to own things, similarly to physical ownership of items such as property. There are four main types of IP (Figure 2.1):

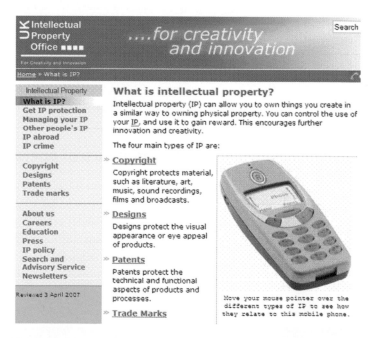

Figure 2.1 Intellectual Property Office
www.ipo.gov.uk/whatis.htm

- **copyright** – protects materials produced such as drawings, text, music, recordings, films and broadcasts
- **designs** – protect the visual appearance of items
- **patents** – protect the technical and functional aspects of items

- **trademarks** – protect signs, labels or other distinguishing features that set aside one item from another.

---

### Activity 2.4

1. Have a look at the UK Intellectual Property Office website and view the information on each of the IP areas mentioned.
2. From the website identify what IP protection is available.

---

## Understand how to access sources of finance

Several internal and external sources of finance are available to all types of businesses, small or large. These sources can be self-generated from savings, investments or profits taken from the business, while other sources may rely on the donations of third party external agencies such as banks and building societies, in the form of loans or mortgages or other investment stakeholders in terms of providing hire purchase (HP) or a lease provision.

### Internal and external sources

Internal sources of finance include money that is invested by the owner from savings, etc., or from capital.

External sources of finance include money that is borrowed or loaned from banks in the form of overdrafts (Figure 2.2), business loans, commercial mortgages, venture capital, HP, leasing, factoring or share issues.

Figure 2.2  HSBC business services
http://www.hsbc.co.uk/1/2/

### Activity 2.5

Complete the table below by providing a definition and example of the following external sources of capital:

| External capital source | Definition | Example of where this can be obtained |
|---|---|---|
| Overdraft | | |
| Business loan | | |
| Commercial mortgage | | |
| Venture capital | | |
| HP | | |
| Leasing | | |

## Be able to interpret financial statements

Financial statements are vital to any organization. The need for written, auditable evidence that shows the historic, current and projected financial status of an organization is essential. Organizations need to know about every aspect of how they are performing as this will impact on the resources that they have and possibly require in the future, which products and services they buy and sell, and how they buy and sell them, and how they compete in the market; and also it provides them with a projection of how they can function in the future.

### Costs and budgets

Costs managed to budget include fixed and variable costs, breakeven, monitoring budgets and variances. Income increased to budget, bids to secure future resources, e.g. capital grants and investment. Provision of appropriate liquidity/working capital and adequate emergency reserves.

#### Cost–benefit analysis

A cost–benefit analysis can be used to justify the cost of a proposal by offsetting this against the benefits of the cost. Regardless of whether you have a virtual or real end-user, costs may need to be considered.

A cost–benefit analysis provides an overview of the costs involved with a certain project and maps these against the attributed benefits.

In theory, every cost listed should have at least one benefit marked against it to justify the expense. The costs can be broken down into two areas:

- tangible costs
- intangible costs.

Tangible costs are costs that can be assigned to physical items such as new hardware, software or additional office equipment. Intangible costs are non-physical items such as installation of the hardware or training to use the new software.

A cost–benefit analysis can be produced by drawing up a matrix similar to the one shown in Table 2.2.

**Table 2.2** Cost–benefit analysis

| Tangible costs | £ | Intangible costs | £ | Benefits |
|---|---|---|---|---|
| Three new PCs | 4,700.00 | Installation of PCs | 160.00 | Wider access to the system by users |
| | | | | Reduction in processing activities |
| | | | | Improved customer service due to faster response time |
| New desk-top publishing software | 400.00 | Training on the software | 350.00: 35.00 an hour for 10 hours | Ability to create company documents such as letterheads and logos in-house |
| | | | | Saving of £650.00 per quarter by producing promotional material in-house and not via a marketing agency |
| | 5,100.00 | | 510.00 | |

## Financial statements

Financial statements provide information about the value of an organization. Some organizations, i.e. those that are publicly owned, are obliged to publish periodic financial statements that include a balance sheet, an income statement (profit and loss account) and a cash-flow statement/analysis.

A balance sheet represents the assets, liabilities and owners' equity of the company at a specific point in time (Table 2.3). A balance sheet has to follow a set formula:

$$\text{Assets} = \text{Liabilities} + \text{Shareholders' equity}$$

Each of these segments will have various accounts embedded within them, such as:

Assets = cash, inventory and property
Liabilities = accounts payable or long-term debt
Equity = shareholders' investments and retained earnings.

A profit and loss account presents details about the earnings achieved for a certain period (Table 2.4).

The statement of cash flows presents cash receipts and payments for various operating, investing or financial activities and provides definitions of each category (Table 2.5).

## Basic ratios

A number of ratios can be used by organizations to determine their financial state in terms of:

- solvency
- profitability
- performance.

Table 2.3 Sample balance sheet: Tesco plc

**TESCO PLC**
**GROUP BALANCE SHEET (Unaudited)**

| | 12 Aug 2000 £m | 26 Feb 2000 £m |
|---|---|---|
| **Fixed assets** | | |
| Intangible assets | 132 | 136 |
| Tangible assets | 8,561 | 8,140 |
| Investments | 74 | 79 |
| Investments in joint ventures | 193 | 172 |
| | 8,960 | 8,527 |
| **Current assets** | | |
| Stocks | 780 | 744 |
| Debtors | 233 | 252 |
| Investments | 195 | 258 |
| Cash at bank and in hand | 116 | 88 |
| | 1,324 | 1,342 |
| **Creditors: falling due within one year** | (3,474) | (3,487) |
| **Net Current Liabilities** | (2,150) | (2,145) |
| **Total Assets Less Current Liabilities** | 6,810 | 6,382 |
| **Creditors: falling due after more than one year** | (1,754) | (1,565) |
| **Provisions for liabilities and charges** | (19) | (19) |
| **Total Net Assets** | 5,037 | 4,798 |
| **Capital and Reserves** | | |
| Called up share capital | 343 | 341 |
| Share premium account | 1,698 | 1,650 |
| Other reserves | 40 | 40 |
| Profit and loss account | 2,924 | 2,738 |
| **Equity Shareholders' Funds** | 5,005 | 4,769 |
| Minority interest | 32 | 29 |
| **Total Capital Employed** | 5,037 | 4,798 |

http://www.tesco.com/recruitment/html/careers/complnfo/statements.htm

The solvency ratio is used to measure an organization's ability to meet its long-term obligations and can be calculated as:

$$\text{Solvency ratio} = \frac{\text{After-tax net profit} + \text{Depreciation}}{\text{Long-term liabilities} + \text{Short-term liabilities}}$$

Two ratios can be used to determine the solvency of a business: the current ratio and the acid test ratio.

The current ratio provides a comparison of current assets and current liabilities that measures the ability to pay current debts:

$$\text{Current ratio:} \frac{\text{Current assets}}{\text{Current liabilities}}$$

Table 2.4 Sample profit and loss account

| TESCO PLC GROUP PROFIT AND LOSS ACCOUNT (Unaudited) 24 weeks ended 12 August 2000 | 2000 £m | 1999 £m | Increase % |
|---|---|---|---|
| **Sales at net selling prices** | 10,084 | 9,112 | +10.7 |
| **Turnover excluding value added tax** | 9,302 | 8,423 | +10.4 |
| – Operating expenses | (8,821) | (7,988) | |
| – Employee profit sharing | (19) | (18) | |
| – Integration costs | – | (3) | |
| – Goodwill amortisation | (3) | (3) | |
| **Operating profit** | 459 | 411 | +11.7 |
| Share of operating profit of joint ventures | 10 | 5 | |
| Net (loss)/profit on disposal of fixed assets | (4) | 4 | |
| **Profit on ordinary activities before interest** | 465 | 420 | +10.7 |
| Net interest payable | (50) | (39) | |
| **Profit on ordinary activities before taxation** | 415 | 381 | +8.9 |
| **Profit before integration costs, net (loss)/profit on disposal of fixed assets and goodwill amortisation** | 422 | 383 | +10.2 |
| Goodwill amortisation | (3) | (3) | |
| Integration costs | – | (3) | |
| Net (loss)/profit on disposal of fixed assets | (4) | 4 | |
| Taxation | (114) | (107) | |
| **Profit on ordinary activities after taxation** | 301 | 274 | +9.9 |
| Minority interest | – | – | |
| **Profit for the financial period** | 301 | 274 | +9.9 |
| Dividends | (101) | (90) | |
| **Retained profit for the financial period** | 200 | 184 | |
| | **Pence** | **Pence** | |
| **Earnings per share** | 4.45 | 4.10 | |
| **Diluted earnings per share** | 4.38 | 4.04 | |
| **Adjusted diluted earnings per share** | 4.48 | 4.07 | +10.1 |
| (excluding integration costs, net (loss)/profit on disposal of fixed assets and goodwill amortisation) | | | |
| **Dividend per share** | 1.48 | 1.34 | +10. |

http://www.tesco.com/recruitment/html/careers/compInfo/statements.htm

The acid test ratio indicates whether or not a company has adequate short-term assets to cover its liabilities without selling any inventory:

$$\text{Acid test ratio:} \frac{(\text{Cash} + \text{Accounts receivable} + \text{Short-term investments})}{\text{Current liabilities}}$$

The profitability of a business can be determined by gross and net profit percentages and by examining the return on capital employed.

To determine the performance of a business, stock turnover, the debtors' collection period and asset turnover can be examined.

Table 2.5  Statement of cash flow

| TESCO PLC GROUP CASH FLOW STATEMENT (Unaudited) | | |
|---|---|---|
| 24 weeks ended 12 August 2000 | 2000 £m | 1999 £m |
| Net cash inflow from operating activities | 674 | 734 |
| Returns on investments and servicing of finance | | |
| Interest received | 33 | 30 |
| Interest paid | (65) | (66) |
| Net cash outflow from returns on investments and servicing of finance | (32) | (36) |
| Taxation | (114) | (3) |
| Capital expenditure and financial investment | | |
| Payments to acquire tangible fixed assets | (760) | (543) |
| Receipts from sale of tangible fixed assets | 32 | 48 |
| Purchase of own shares | – | (14) |
| Net cash outflow from capital expenditure and financial investment | (728) | (509) |
| Acquisitions | | |
| Purchase of subsidiary undertakings | – | (61) |
| Investments in joint ventures | (23) | (11) |
| Net cash outflow from acquisitions | (23) | (72) |
| Equity dividends paid | (195) | (177) |
| Cash outflow before use of liquid resources and financing | (418) | (63) |
| Management of liquid resources | | |
| Decrease/(increase) in short term deposits | 63 | (22) |
| Financing | | |
| Ordinary shares issued for cash | 33 | 9 |
| New finance leases | 13 | – |
| Increase in other loans | 343 | 90 |
| Capital element of finance leases repaid | (12) | (3) |
| Net cash inflow from financing | 377 | 96 |
| Increase in cash in the period | 22 | 11 |

http://www.tesco.com/recruitment/html/careers/complnfo/statements.htm

## Assessment activities

| Grading criteria | Content | Suggested activity |
|---|---|---|
| **Pass** | | |
| P1 | Describe how a selected business manages its existing human resources. | Carry out research on a selected business (possibly a place of work, or an organization that is known to you and that you have access to). Produce a presentation that describes how they manage their existing human resources. |
| P2 | Describe the main physical and technological resources that need to be considered in the running of a selected organization. | Extend your presentation to include slides on physical and technological resources that should be considered in the running of the selected organization chosen for P1. |
| P3 | Describe where sources of finance can be obtained for starting up a selected business. | Produce an information leaflet designed to help new business owners decide how to finance their business. The leaflet should explore potential financial sources. |
| P4 | Give the reasons why costs and budgets need to be controlled. | In conjunction with P3 produce an information leaflet that identifies reasons why costs and budgets need to be controlled. |
| P5 | Interpret the contents of a given profit and loss account and balance sheet. | Carry out research on a given organization that illustrates its financial state. The organizational data should include a profit and loss account and a balance sheet for you to interpret. The interpretation could be in the form of a short written analysis. |
| P6 | Illustrate the financial state of a given business by showing examples of accounting ratios. | Demonstrate examples of accounting ratios based on the financial state of the business provided in P5. |
| **Merit** | | |
| M1 | Assess how managing human, physical and technological resources can improve the performance of a selected organization. | In conjunction with P2 you should provide an assessment (8–10 slides) on how managing human, physical and technological resources can improve the performance of the organization chosen. |
| M2 | Analyse the reasons why costs and budgets need to be controlled and explain in detail problems that can arise if they are left unmonitored. | The information sheet (P3) could be extended to analyse the reasons why costs and budgets need to be controlled and a detailed explanation should be provided that details what problems could arise if they are left unmonitored. |
| M3 | Interpret the contents of a given profit and loss account and balance sheet and explain in detail how accounting ratios can be used to monitor the financial state of the organization. | In conjunction with P5 and P6 interpret the contents of the profit and loss account and balance sheet and explain in detail how accounting ratios can be used. You should also be able to monitor the financial state of the organization. |
| **Distinction** | | |
| D1 | Evaluate how managing resources and controlling budgets can improve the performance of a business. | Produce a short report that evaluates how managing resources and controlling budgets can improve business performance. |
| D2 | Evaluate the adequacy of accounting ratios as a means of monitoring business health in a selected organization, using examples. | In conjunction with D1, evaluate the adequacy of accounting ratios as a means of monitoring business health in a selected organization (possibly the one chosen for P1), providing examples. |

CHAPTER 2

Courtesy of iStockphoto, JLFCapture, Image# 4076557

Marketing activities are undertaken and approached in different ways depending upon the size and nature of an organisation. Some organisations place marketing at the core of their business activities and use this to drive promotions and sales.

There are a number of ways that marketing can be deployed within an organisation. Some organisations may choose paper-based methods of advertising, others might select electronic methods through the use of websites, television, radio and other e-media formats. Visual marketing, using different ways and techniques to sell products and services is also very effective in terms of reaching out to a specific audience.

Everyday and almost everywhere you go the consumer is bombarded by products and services, the power of marketing can influence what you buy, which brand, when you buy it, where you buy it from and at what price.

# Chapter 3

# Introduction to Marketing

Marketing is pivotal to any organization. Understanding what market you are in, your customer base, your competitors and trends is vital to the success of an organization.

Customer-focused marketing ensures that your customer is at the heart of any research and promotional activity, and drives any marketing strategy to ensure that the products or services are targeted to the right audience, at the right time and for the right price.

This chapter will provide students with an underpinning knowledge of marketing concepts. The following learning outcomes will be addressed:

- Understand the concept and principles of marketing and their application in the business environment.
- Know how and why marketing research is conducted by organizations.
- Understand how marketing information is used by organizations.
- Understand how marketing techniques are used to increase demand for products (goods and services).

## Understand the concept and principles of marketing and their application in the business environment

Marketing is a crucial element to any business organization and as such strategies that are formulated to ensure the successful launch and longevity of a product or service within the market are based on various marketing principles and the marketing mix. A number of limitations and constraints can also contribute to the success or failure of a product or service, all of which will be explored in the following sections.

### Principles of marketing

Marketing is a critical element of any organization. The concept of marketing is to generate awareness and interest and engage in the process of research, promotion, selling and distribution of products or services. Although the concept of marketing is very customer driven now, this has evolved over the years. In the past organizations were orientated more towards 'production' and reducing costs, and the 'product' in terms of improving the quality, features or accessibility. Sales orientation and making a product aimed at a particular target audience was another driving factor. However, the focus and strategic aim of the large majority of organizations today is to be 'market' orientated and to put the customer at the centre and heart of the business.

Marketing can mean different things to different people, depending on the product or service type, organization type, available resources and budget.

The Chartered Institute of Marketing [Chartered Institute for Marketing (CIM) http://www.cim.co.uk] defines marketing as:

> The management process responsible for identifying, anticipating and satisfying customer requirements profitability

whereas the American Marketing Association [American Marketing Association http://www.marketingpower.com] approved the following definition in October 2007:

> Marketing is the activity, set of institutions, and processes for creating, communicating, delivering, and exchanging offerings that have value for customers, clients, partners, and society at large.

However marketing is defined, at the core of this activity is the customer and meeting the demands and needs of the customer.

The principles and activities of marketing vary depending on the product or service, and the type and size of organization, as mentioned before. Marketing activities are driven by the orientation of the company, in terms of their focus being production, product, sales or market orientated. As a result this orientation will provide the key driver to any marketing activity undertaken.

## Marketing objectives

To support the application of marketing within an organization, there needs to be a formal framework that is built around clear marketing aims and objectives. Any objectives set should also be SMART, defined in terms of being:

**S**pecific
**M**easurable
**A**chievable
**R**ealistic
**T**ime-framed.

Working within the boundaries of SMART objectives should enable an organization to outline plans for the future and to ensure that the implementation of these plans is relatively seamless.

---

**Activity 3.1**

An example of a SMART marketing target might be:

To increase the current market share of supplying organic fresh produce to supermarkets by 25% within the next two years.

1. Using the example, provide four other appropriate marketing SMART targets.

---

## Functions

The functions of marketing can be diverse, depending on the type and size of an organization. For some organizations marketing is critical to the success of a given product or service; the need to promote a brand and generate awareness is one of the functions that marketing can perform. Marketing can be used to pilot or test a new product or service, promotional activities being used to analyse customer reactions. Marketing can also be used as an evaluative tool to assess the success of a product by gathering customer feedback and opinions.

## Organizational and marketing objectives

Organizational and marketing objectives should be intertwined and support each other. If an organization had set objectives that involved growth in the market, increasing a product range over the next year and increasing profits by 15%, the marketing objectives should be tools that can be used to achieve this.

## Use of marketing principles

Marketing principles will also vary between different organizational sectors. A public organization may use marketing to generate an awareness about a given social or environmental issue, for example. A private organization may adopt a very aggressive marketing strategy to ensure that they maintain their position within the market, or to gain a greater market share. Voluntary organizations may apply marketing strategies to campaign for a particular cause, such as vaccinations in developing countries or the NSPCC 'full-stop' appeal.

Marketing principles will also depend on the customer, third party or shareholder, who the marketing is aimed at, and the perceived impact. Other factors are planning, control and evaluation, and the development of e-marketing.

## Marketing mix

Successful marketing strategies will be based on what is referred to as the 'marketing mix'. The marketing mix is based on the 4 Ps, as shown in Figure 3.1.

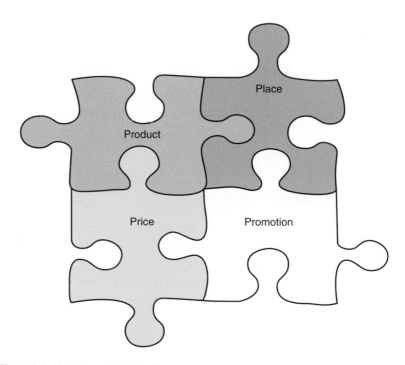

**Figure 3.1** Marketing mix: the four Ps

The **product** is the physical product/stock item or service that you are selling to your target audience.

The **place** could be the extent or boundaries, geographically or virtually, of your marketplace and target audience. For example, you could have a small shop or outlet based in a particular location that would serve a target audience possibly within a certain mile radius, or you could have an online provision that serves a global community.

The **price** is the selling price for your product or service. The unit price will take into consideration the overheads and costs associated with researching, physically making/manufacturing, promoting and distributing the product or service.

**Promotion** is a critical part of any product or service. Promotion will involve creating awareness and selling the product or service to a particular target audience. This can be in the form of:

- **paper-based** – leaflets, posters, brochures and booklets, questionnaires
- **electronic media** – television advertising, website promotion, text messaging
- **radio advertising**
- **direct promotion** – cold calling, taster sessions.

## Limitations and constraints on marketing

Numerous legal, data and voluntary constraints are placed on marketing and organizations that partake in marketing activities. Marketing activities are also constrained by public opinion, pressure from certain groups and consumerism.

Legal protection is available to cover consumers and protect them from certain promotional and selling activities. These consumer laws include:

- Sale of Goods Act 1979
- Trade Descriptions Act 1968
- Consumer Credit Act 1974
- Data Protection Act 1998.

### Sale of Goods Act 1979

The Sale of Goods Act regulates contracts in which goods and services are bought and sold. Under this Act traders must comply and sell items that are as described and of satisfactory quality. If consumers find that items purchased do not match these criteria they can reject them and ask for a refund, or they can request a replacement or a repair or some other form of compensation.

### Trade Descriptions Act 1968

The Trade Descriptions Act offers protection to consumers and traders. There is a number of areas to the Act that includes the following within the meaning of 'trade description':

- **Quantity or size** – including the length, width, height, area, volume, capacity, weight and number of an item.
- **Composition** – what the item is made from.
- **Fitness for purpose, strength, performance, behaviour or accuracy** – the item must be suitable for the intention for which it is bought and adhere to the description given.
- **Any physical characteristics that have not been previously mentioned** – any description not previously referred to would amount to a trade description, i.e. method of manufacture, production, processing or reconditioning.
- **Testing by any person and results thereof** – including any statement that a car has had its mileage independently checked.

- **Approval by any person or conformity with a type approved by any person** – may include a statement that an item complies with a British Standard.
- **Place or date of manufacture, production, processing or reconditioning**.
- **Person who manufactured, produced, processed or reconditioned the item** – including things such as brand names on an item. This section is often used to deal with counterfeit goods.
- **Other history** – including previous ownership or use.

### Consumer Credit Act 2006

The Consumer Credit Act provides an arena for a fairer, clearer and more competitive market for consumer credit.

#### Key dates

On 6 April 2007, the remit of the Financial Ombudsman Service (FOS) was extended to cover consumer credit and the Unfair Relationships Test was introduced for new agreements.

On 6 April 2008, the Office of Fair Trading's (OFT's) new strengthened licensing regime was introduced, the Consumer Credit Appeals Tribunal (for appeals against OFT licensing decisions) was established, the financial limit (of £25,000) was removed so that all new credit agreements (unless specifically exempt) regardless of value are regulated, and the Unfair Relationships Test was extended to all existing credit agreements.

Since 1 October 2008, lenders have been required to provide borrowers with much more information about their accounts, such as an annual statement and regular notices when consumers fall into arrears or incur a default sum, and debt administration service providers and credit information (repair) service providers need a consumer credit licence as these services are now regulated by the OFT.

### Data Protection Act 1998

The Data Protection Act applies to the processing of data and information by any source, either electronic or paper based. The Act places obligations on people who collect process and store personal records and data about consumers or customers. The Act is based on a set of principles which binds a user or an organization into following a set of procedures offering assurances that data is kept secure.

#### Main principles

1. Personal data shall be processed fairly and lawfully and, in particular, shall not be processed unless:
   - at least one of the conditions in Schedule 2 of the 1998 Act is met and
   - in the case of sensitive personal data, at least one of the conditions in Schedule 3 of the 1998 Act is also met.

2. Personal data shall be obtained only for one or more specified and lawful purposes, and shall not be further processed in any manner incompatible with that purpose or those purposes.
3. Personal data shall be adequate, relevant, and not excessive in relation to the purpose or purposes for which they are processed.
4. Personal data shall be accurate and, where necessary, kept up to date.
5. Personal data processed for any purpose or purposes shall not be kept for longer than is necessary for that purpose or those purposes.
6. Personal data shall be processed in accordance with the rights of data subjects under this Act.
7. Appropriate technical and organizational measures shall be taken against unauthorised or unlawful processing of personal data and against accidental loss or destruction of, or damage to, personal data.
8. Personal data shall not be transferred to a country or territory outside the EEA (European Economic Area) unless that country or territory ensures an adequate level of protection for the rights and freedoms of data subjects in relation to the processing of personal data.

The Act gives rights to individuals in respect of personal data held about them by data controllers. These include the rights:

- to make subject access requests about the nature of the information and to discover to whom it has been disclosed
- to prevent processing likely to cause damage or distress
- to prevent processing for the purposes of direct marketing
- to be informed about the mechanics of any automated decision-taking process that will significantly affect them
- not to have significant decisions that affect them made solely by an automated decision-taking process
- to take action for compensation if they suffer damage by any contravention of the Act by the data controller
- to take action to rectify, block, erase or destroy inaccurate data and
- to request the commissioner to make an assessment as to whether any provision of the Act has been contravened.

The Act does provide wide exemptions for journalistic, artistic or literary purposes that would otherwise be in breach of the law.

### Useful definitions
- **Personal data** – Information about living, identifiable individuals. Personal data does not have to be particularly sensitive information and can be as little as name and address.
- **Data users** – Those who control the contents, and use of, a collection of personal data. They can be any type of company or organization, large or small, within the public or private sector. A data user can also be a sole trader, a partnership or an individual. A data user need not necessarily own a computer.
- **Data subjects** – The individuals to whom the personal data relates.
- **Automatically processed** – Processed by computer or other technology such as documents image-processing systems.

### Role of the Data Protection Commissioner

The Commissioner is an independent supervisory authority and has an international role as well as a national one. Primarily the Commissioner is responsible for ensuring that the Data Protection legislation is enforced.

In the UK, the Commission has a range of duties including:

- promotion of good information handling
- encouraging codes of practice for data controllers.

To carry out these duties the Commissioner maintains a public register of data controllers. Each register entry contains details about the controller such as their name and address and a description of the processing of the personal data to be carried out.

### Registering entries

All users, with a few exceptions, have to register an entry or entries giving their name, address and broad descriptions of:

- those about whom personal data is held
- the items of data held
- the purposes for which the data is used
- the sources from which the information may be disclosed, i.e. shown or passed to
- any overseas countries or territories to which the data may be transferred.

Voluntary constraints that are placed on marketing activities can include compliance with the Code of Advertising Practice and the Advertising Standards Authority (Figure 3.2).

**Figure 3.2** Advertising Standards Authority
http://www.asa.org.uk/asa/

Pressure groups and consumerism will also dictate to a certain extent what is deemed to be acceptable in terms of advertising, through to packaging and pricing of products and services, and even the language used in any campaigns.

---

**Activity 3.2**

1. Provide two examples of cases where pressure groups or consumerism have dictated what is deemed to be 'acceptable' in terms of advertising.

---

## Know how and why marketing research is conducted by organizations

Market research is used to obtain data about public opinion that can be used to improve the quality, set the pricing and make more informed decisions about a particular product or service.

### Marketing research

Marketing research can be segregated in terms of the type and quality of the research conducted. For example, research can be defined as being qualitative or quantitative depending on whether it is information or data rich. Research can also be classified as being primary or secondary based on the method, purpose and source of data collected.

#### Qualitative and quantitative research

Quantitative research and information is based on facts and statistics, key information used for finance, planning and modelling, etc. Examples of this type of information are sales figures, control measurements and test data for an experiment. See Figure 3.3 for an example of this. There is a great need for quantitative research, especially if you work within mathematical, scientific, medical or logic-orientated environments where calculations and experimentation are required.

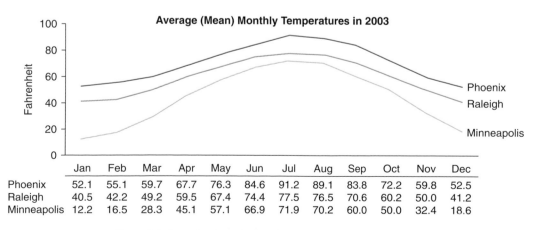

| | Jan | Feb | Mar | Apr | May | Jun | Jul | Aug | Sep | Oct | Nov | Dec |
|---|---|---|---|---|---|---|---|---|---|---|---|---|
| Phoenix | 52.1 | 55.1 | 59.7 | 67.7 | 76.3 | 84.6 | 91.2 | 89.1 | 83.8 | 72.2 | 59.8 | 52.5 |
| Raleigh | 40.5 | 42.2 | 49.2 | 59.5 | 67.4 | 74.4 | 77.5 | 76.5 | 70.6 | 60.2 | 50.0 | 41.2 |
| Minneapolis | 12.2 | 16.5 | 28.3 | 45.1 | 57.1 | 66.9 | 71.9 | 70.2 | 60.0 | 50.0 | 32.4 | 18.6 |

**Figure 3.3** Example of quantitative information
Source: http://www.perceptualedge.com/example2.php

Qualitative research can be used to make detailed and descriptive decisions that could affect a product launch or new product design. The basis of qualitative information is to probe and question to gain an understanding of the subject matter, for example a survey conducted about customer opinions on a certain product. Statistics might have been generated to identify retention and achievement rates as a percentage based on similar conditions across the country. However, the qualitative aspect may illustrate through the use of a course feedback form which modules students enjoyed, and how they felt about assessments, support and resource issues that may have helped them to achieve, etc.

The best ways to extract qualitative information are through interviewing, questionnaires (Figure 3.4), feedback forms and surveys.

| **Customer survey – software packages** | |
|---|---|
| Name: | Age (please tick the appropriate box) |
| Address: | 18 or below years  19–35 years  36–50 years<br>51–64 years          65+ years |
| Please tick the types of software package purchased new in the last twelve months | Please indicate if you have just upgraded an existing software package in the last twelve months (added to the existing software) |
| | Upgraded |
| Word processing | Yes/No |
| Spreadsheet | Yes/No |
| Database | Yes/No |
| Graphics | Yes/No |
| Presentation | Yes/No |
| Utility | Yes/No |
| Multimedia | Yes/No |
| Specialist | Yes/No |
| Other<br>(please specify) | Yes/No |
| What is the maximum you would pay for each software package?<br><br>Word processing  £        Spreadsheet  £        Database  £<br>Graphics            £        Presentation  £        Utility      £<br>Multimedia        £        Specialist      £        Other      £ | |
| Where do you purchase your software, e.g. mail order, retail shop, direct from manufacturer, etc., and why? | |
| What features do you look for when purchasing software, e.g. user friendly, compatibility with existing software, etc. | |
| What has influenced your decision to purchase software from us in the past (please rate from 1–5) 1 = lowest 5 = highest<br><br>Competitive prices<br>Good quality service<br>Fast despatch<br>Regular updates and new product information<br>Friendly staff | |
| How could we improve our service to you? | |

Figure 3.4  Sample data capture document: customer survey

---

**Activity** 3.3

1. Design a questionnaire to satisfy one of the following:
   - survey types and frequency of magazine purchases
   - investigate the best film of the year.

For each questionnaire include a mixture of fifteen qualitative and quantitative questions. When the questionnaire design is complete, print off or e-mail ten copies to people within the group and get them to complete them.

2. From the information gathered produce two charts and one graph based on the quantitative information and provide a short summary detailing the findings of the qualitative information.

3. Identify which was the easier of the two sets of information to produce: qualitative or quantitative. Why do you think this is?

---

## Primary research

Primary research can be gathered internally from within an organization or externally through direct marketing activities. Primary research can be used to determine all aspects of a new product or service launch or campaign from how the item is packaged through to the price, distribution method and physical location.

The purpose of primary research is to gather first hand information from customers and consumers and other interested parties (competitors, etc.) about a specific product or service and possibly the environment in which it is marketed and sold. Therefore it is crucial that any information gathered is:

- accessible – available to the right people
- fit for purpose – useful
- valid and reliable
- accurate and timely.

Marketing research allows organizations to examine a range of characteristics about their customers to:

- sell more products/services
- reduce or increase prices
- target sales
- change the products or service range
- introduce a new product or service.

The ways in which organizations can gather primary information vary, as shown in Figure 3.5. Marketing information, as shown in Figure 3.5, can be generated from a number of sources. One of the easiest and most widely used ways to gather consumer information is through the use of a loyalty scheme or card that entices consumers with various benefits. Many organizations use this scheme as a form of primary research (Table 3.1), especially in supermarkets and large retail stores to determine:

- how much consumers spend
- how often consumers shop
- what consumers buy

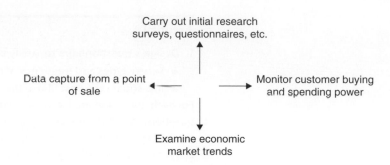

Figure 3.5 Ways of collecting marketing information

Table 3.1 Loyalty schemes offered by organizations

| Organization | Scheme | Reward |
| --- | --- | --- |
| Tesco | Clubcard | A range of leisure pursuits/discount vouchers off shopping |
| Sainsbury's/BP/ Debenhams | Nectar | Flights/holiday vouchers/store discounts |
| Boots | Advantage Card | Discounts/health and well-being products and services |
| WH Smith | Clubcard | Discounts off products in store |
| Homebase | Spend and Save | Discounts off products in store |

- trends in consumer buying patterns
- information about consumers' lifestyles.

## Secondary research

Secondary research is also important in the marketing process because it can be made available from third party sources that specialize in gathering information about specific types of consumers and their buying patterns. In addition, some organizations have been set up purely for the purpose of buying and selling information. The benefits of buying information off a third party source include the fact that it can be tailored to meet a specific need, for example information about the buying patterns of 16–25 year olds in the last six months in terms of games console purchases.

The information purchased will also be up to date and current; therefore, any further analysis and use of this information will be relevant and appropriate.

### Case Study 3.1

### Credit agencies: the value of information

All information can be deemed to be of value to someone; however, some forms of information classified as secondary information can be purchased by organizations to help support a range of business activities, including marketing. An example of information being used in this way is credit agency companies (Figure 3.6).

Credit agencies are used by a number of retail, banking and financial institutes to check the credit rating of consumers who want to take out a loan, mortgage or credit

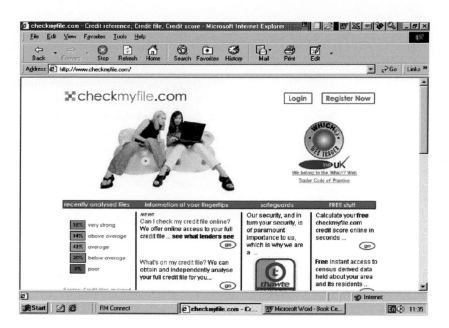

**Figure 3.6** Sample credit agency: Checkmyfile.com
http://www.checkmyfile.com/

facilities on goods and services. An organization can assess a consumer's ability to honour payments by carrying out a credit check. The process of checking a credit rating is based on a scoring system where historical data, personal finances and stability are assessed. The more points accumulated the greater the chance of being approved for credit.

Credit agencies gather this information from a variety of sources such as electoral rolls, public opinion surveys, other organizations and agencies. This information is then sold on to organizations requiring credit check services. In conjunction, updated information about credit check status, approvals and rejections are passed back to the credit agency.

Members of the public can access information on their own credit history, but will incur an administrative charge to do so.

The credit agency case study is one example of how secondary information can be used. Organizations pay for the information on applicants for loans and credits that can then support their own marketing activities, for example are they selling their product or service to the right consumer group or do they need to rethink their target audience?

## Activity 3.4

**Using a variety of sources identify at least three credit agencies and carry out the following:**

1. **What sort of information is gathered about consumers/applicants?**
2. **Provide five examples of the types of organization that may use the services of these agencies.**
3. **How much does it cost to access your own credit file?**

**CHAPTER 3**

### Activity 3.5

**Discuss in groups of three or four:**

1.  Do you think that information should be available to be 'bought' and 'sold'? If yes, under what conditions?
2.  Do you think that it is ethical for an organization such as a credit agency to collect information on individuals?

A range of other organizations provides secondary research and information for organizations to support them with their marketing activities. These include Mintel, Bradstreet and Datastream:

http://www.mintel.com/

http://www.dnb.com/us/

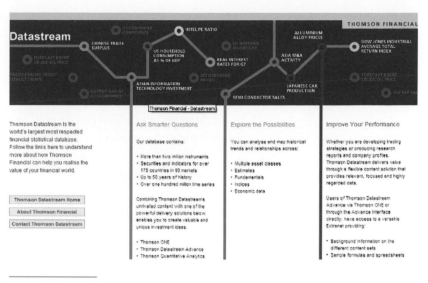

http://www.datastream.com/

Information can also be gathered from government polls and statistics, for example the Social Trends and Family Expenditure Survey (Figure 3.7).

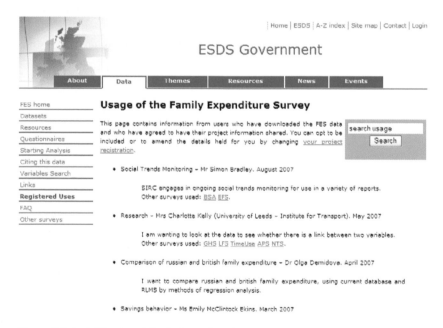

**Figure 3.7** Social Trends and Family Expenditure Survey
http://www.esds.ac.uk/government/fes/usage/

## Understand how marketing information is used by organizations

Marketing information can be used for a variety of different purposes in organizations. Information can be used to identify consumer preferences and gaps in the market and to possibly identify niche markets. It can also be used to set business objectives and to help to understand and appreciate consumer preferences, different lifestyles, competitors' activities the general environment in which a particular product or service operates.

## Using marketing research

Marketing research can be used to support the growth and success of a particular product or service to the extent that it can even be used to ascertain whether or not a product or service reaches the market and its intended audience.

Marketing research can be used to identify certain themes that occur within a specific area, for example:

- what products people buy and why
- when consumers buy products – time of day, etc.
- how much consumers will pay
- age, socioeconomic grouping, etc.

If common themes are identified marketing can then be targeted even further to embrace consumer preferences and lifestyles.

Understanding competitors and their marketing strategies is also important in terms of the growth and success of an organization's own marketing activities. A good example of this can be seen with supermarkets and promotions on offer such as BOGOF (buy one get one free) and special promotion items that have been discounted in some way or featured in a particular area in an aisle. Another example is the product ranges offered by supermarkets, each having three distinct product sectors with a budget and premium range as extremes (Table 3.2).

Table 3.2 Product ranges offered by supermarkets

| Supermarket | Budget range | Premium range |
|---|---|---|
| Tesco | Value | Finest |
| Sainsbury's | Basic | Taste the Difference |
| Asda | Smartprice | Extra Special |

## Case Study 3.2

### Premium brands boost Sainsbury's as Britain gets the taste for fine foods

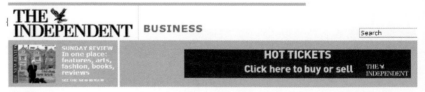

J Sainsbury became the latest group to hail shoppers' new-found appetite for top-quality food as a key factor behind its 5 per cent leap in like-for-like sales over the past three months.

Mums on a budget should look away now: Britain is in the grip of a speciality food boom that is feeding straight through to the bottom lines of the country's top supermarket groups after one of the best Christmases for the grocery industry in years.

Already this week, WM Morrison, Waitrose and, to a lesser extent, Marks & Spencer have benefited from a 'flight-to-quality' by shoppers the length and breadth of the land.

Sales of Taste the Difference, Sainsbury's premium own label range, grew by 20 per cent during the period on the back of strong sales of just about everything from vintage champagne to jumbo king prawns. Morrisons' equivalent brand, The Best, grew by 40 per cent over Christmas, while Asda's Extra Special grew even faster still, at 61 per cent.

Figures from the IGD, the grocery think-tank, predict that the value of retailers' premium private labels will almost double during the next four years, to a shade under £9 bn. A separate report from Datamonitor, the research group, today predicts that retail sales of speciality foods and drinks will hit £4.2 bn by 2011, up from £3.6 bn last year.

All of the top grocers have relaunched their premium lines in the past few months. And that includes Asda, a chain hardly renowned as a magnet for the high-spending shopper. Andy Bond, its chief executive, has said: 'We see a shift in our range towards more premium lines.' He wants to triple the amount of revenues that come from its 'best' lines to 10 per cent over the next three years.

Patrick Mitchell-Fox, IGD's senior business analyst, said: 'Retailers that are not involved in the premium market are now the exception rather than the norm.'

When Tesco issues its trading update on Tuesday, it will highlight how customers cannot get enough of its Finest ranges, which it recently extended to non-food products from cashmere jumpers to saucepans. Sales of its Finest lines are growing at four times the rate of the rest of its business.

In fact, much as they would hate to admit it, all of Tesco's rivals probably owe it a big 'thank you' for inventing the premium ready meals genre back in 1998. Tesco was the first of the big grocers to realise that it could appeal to more customers if it segmented the types of food it sold. The Finest brand is now worth £1 bn a year – that's 3 per cent of its annual UK sales.

Sainsbury's is doing better still, relatively speaking. Taste the Difference is knocking at the door of £1 bn in sales, which is about 17 per cent of its top line. As Justin King, its chief executive, never tires of reminding: 'Sainsbury's was founded on the principles of quality food. It's what our customers want.'

Sainsbury's is attempting to take a further bite out of the premium market with a new range of 'Super Natural' ready meals that have done away with all of the food nasties still found in even most top-quality lines. The new meals are based around 'superfoods' such as pulses, beans and pomegranates and contain no additives or trans-fats, that culinary bogeyman of the Noughties.

By Susie Mesure, Retail Correspondent

12 January 2007

http://www.independent.co.uk/news/business/analysis-and-features/premium-brands-boost-sainsburys-as-britain-gets-the-taste-for-fine-foods-431818.html

## Activity 3.6

1. **Why do you think that premium brand shopping items have become popular within supermarket chains?**

There is a number of lifestyle and aspiration measures and classifications that can also be used by organizations as a form of marketing research, including:

- ACORN
- MOSAIC
- 4Cs

all of which can be used to segment, target and position products.

---

### Activity 3.7

1. Carry out research to identify what ACORN, MOSAIC and the 4Cs are.
2. Produce a short presentation, comprising ten slides, that introduces each of these lifestyle and aspiration measures.
3. Produce a table that compares and contrasts each of these.

---

## Analytical techniques

Several analytical techniques can be used by organizations to support them in understanding and determining marketing information and how it can be used. These techniques include:

- situation analysis
- SWOT (strengths, weaknesses, opportunities and threats) analysis (Table 3.3)
- PESTLE (political, economic, social, technological, legal and economic).

Table 3.3 Stock control SWOT analysis

| Strengths | Weaknesses |
|---|---|
| • Would reduce the amount of paperwork<br>• Would allow automatic stock ordering<br>• Provides automatic tracking of the stock distribution in the warehouse<br>• Information can be accessed by all stock personnel<br>• Would reduce overheads by 5% | • Initial financial outlay<br>• Training of all warehousing personnel |
| **Opportunities** | **Threats** |
| • Stock system can be integrated into other department systems<br>• Access to stock information by all store personnel<br>• Links into supplier systems | • Incompatibility with other existing systems in the store |

A SWOT analysis:

- **S**trengths
- **W**eaknesses
- **O**pportunities
- **T**hreats

is modelled on four key elements, which together provide a holistic view of a particular proposal or project. For example,

Should the stock control system be upgraded in the warehouse department?

To assess whether or not the proposal would be viable, the strengths should outweigh the weaknesses and the opportunities should exceed the threats.

---

### Activity 3.8

1. Provide a short summary on the use of a SWOT analysis as a tool for planning and decision making.
2. Is there anything else that needs to be considered and potentially added to any of the four sectors?
3. In your opinion, what measures could be taken if the weaknesses did outweigh the strengths and the threats exceeded the opportunities in the given example?
4. Produce a SWOT analysis based on your decision to take and complete a BTEC National qualification.

---

Examination of competitors and competitiveness and developments in the local, national and global marketplace can also help an organization to make more informed decisions about how marketing research can be used most effectively in terms of the decision-making process.

Internally, an organization can also examine the product life cycle and its proposed journey from fruition through to development and release into the market.

## Understand how marketing techniques are used to increase demand for products (goods and services)

Marketing techniques can be used to increase demand for products, by understanding the market you are in and the consumer type that you provide for. Certain techniques can allow you to target your product or service more effectively and specifically.

### Marketing segmentation and targeting

Market segmentation and targeting focuses on offering a particular product or service to a particular sector of society or 'target audience'.

Segmentation of the market could be based on geography, by offering a product or service to a particular region or area. Segmentation could also be based on demography, therefore only targeting a specific age group, for example Saga only markets holidays to a more mature, possibly retired consumer sector (Figure 3.8), whereas Club 18–35 only markets holidays to consumers who fall within this age category (Figure 3.9). Market segmentation can also be based on psychographic or lifestyle options. Therefore some products or services are only aimed at families, the retired sector, business people or students.

CHAPTER 3

CHAPTER 3

Figure 3.8  Demographic based segmentation: Saga
http://www.saga.co.uk/travel/General3/

Figure 3.9  Demographic based segmentation: Club 18–35http://www.club18-30.com/club/

## Activity 3.9

Market segmentation allows you to target certain sectors of customers with certain products. For example, GQ magazine is specifically targeted at men.

1.  Carry out research to identify four other products or services that are targeted specifically at a particular audience.
2.  Provide an example where market segmentation is based on geographical region.
3.  Provide an example of a product that is segmented according to gender.

*Marketing mix*

Ensuring that there is a good balance to the marketing mix is crucial in terms of satisfying the needs of the consumer or specific target audience.

Consumers now have greater choice in terms of accessing certain products, services and brands. With the growth in e-business and e-marketing consumers can pick and choose where they access items and they have greater flexibility in terms of pricing and brands.

E-business and e-marketing have enabled greater individualization to customers; the ease of selecting and purchasing products and services. The Internet has opened up a portal of global trading that has branding.

## Branding

Branding is an important element in the marketing process. Establishing a brand name or logo that is recognizable and even possibly desired will contribute to the success and longevity of a product or service.

Branding is very important because it can influence buyer behaviour. We live in a very brand-driven society where labelling is sometimes as important as, if not more important than, the product or service itself. Some consumers will only buy a product or service of a certain branding despite the fact that it may be more expensive or not as functional or practical.

Branding is evident across a range of target audiences and products/services. The must-have new games console, mobile phone, item of clothing or even car drive consumers towards certain brands that have built up and become established within the market in terms of status and positioning. This makes it easier for an organization that wants to extend its brand and diversify into new markets. Consumers will transfer their loyalty and trust over to a new product or service that falls under the umbrella of the same brand.

## Relationship marketing

Establishing a link, loyalty or relationship with a consumer over a period of time can be classed as 'relationship marketing'. For example, a consumer may purchase a vehicle from a car dealership. During the period of purchase, communications between that consumer and dealership will be established to ensure that any information regarding new vehicle models and special offers is sent to the consumer. In addition, as the vehicle gets older details about aftercare, warranties and servicing will also be exchanged.

Another example is that of a couple expecting a baby. Over a period of time the requirements and needs of that family will change as a baby grows and develops. Relationship marketing could be used to anticipate the needs of the family in terms of dates and child ages and then target various aspects of this. Using the car example, a couple may need to trade in a small car or two-seater for a more family-orientated vehicle. Relationship marketing would anticipate and plan for this occurrence and then directly target this family in a few months' time.

## Planning, control and evaluation processes

Effective marketing revolves around careful planning, control and evaluation processes. By using a set marketing planning process that involves auditing data and information of a historical and current nature, marketing decisions can be made and adapted for the future. A number of objectives can also be set that will link into an organization's overall marketing targets and strategies in the short, medium or long term.

Once a strategy has been developed and set, the implementation of any marketing processes will need to be monitored and evaluated to ensure that all initial objectives have been addressed and that any necessary change is enforced.

## Questions and review

1. What is meant by a SMART target/objective? Provide an example of this.
2. How might marketing strategies change between a public, private and voluntary organization?
3. What is the marketing mix?
4. Identify two limitations and constraints on marketing.
5. Provide examples of both qualitative and quantitative research.
6. Produce a questionnaire or survey that could be used to gather customer opinion about a new range of organic grooming products for both men and women. The questionnaire should contain a mixture of quantitative and qualitative questions.
7. It is important for primary research methods to be 'valid' and 'fit for purpose'. Why do you think that these are important elements?
8. Provide three examples of secondary research sources of information.
9. Organizations use marketing information in a number of ways. Understanding customer preferences is an important element in terms of being able to tailor a particular product or service and target it to a specific audience. In what other ways can marketing information be used?
10. Identify two lifestyle and aspiration classifications.
11. What is meant by PESTLE?
12. On what basis could market segmentation take place? For example, an organization could segment the market in terms of geographical regions.
13. Branding is very important and can influence consumer behaviour. Provide evidence of three organizations that have strong branding.
14. What is meant by the term/concept 'relationship marketing'?
15. Provide a brief overview of the marketing planning, control and evaluation process.

## Assessment activities

| Grading criteria | Content | Suggested activity |
|---|---|---|
| **Pass** | | |
| P1 | Describe the concept and principles of marketing. | Produce a presentation that could be used to introduce people to the concepts and principles of marketing. |
| P2 | Describe how the concept and principles are applied to the marketing of products in two organizations. | Carry out research into two organizations that identifies how the concepts and principles are applied to the marketing of their products. This information could then be incorporated within the presentation for P1. |
| P3 | Describe how marketing research information is used by one of the organizations to understand the behaviour of customers, competitors and market environment. | Produce a short case study brochure that describes how marketing research information is used by one of the organizations identified in P2. Focus upon what the organization does in order to understand the behaviour of customers, competitors and the market environment. |
| P4 | Apply two analytical techniques to a selected product (goods or services) offered by a selected organization. | Use information based on one of the organizations selected for P2, or select a different organization, and apply two analytical techniques to a selected product (goods or services). |
| P5 | Describe how marketing techniques are used by one organization to increase demand for a selected product (goods or services). | The case study brochure (P3) could also describe how marketing techniques are used by this organization to increase demand for a selected product (goods or services). |
| **Merit** | | |
| M1 | Compare the effectiveness of the concepts and principles applied to the marketing of products by the two chosen organizations. | In conjunction with P1, compare the effectiveness of the concepts and principles applied to the marketing of products by the two chosen organizations. The evidence should be based on an additional 8–10 slides and an accompanying hand-out. |
| M2 | Compare the analytical techniques used in supporting the marketing decisions of a selected business or product. | In conjunction with P4 the analytical techniques used should be compared in terms of how they support the marketing decisions of a selected business or product. |
| M3 | Explain the marketing techniques used by a selected organization and analyse why these techniques might have been chosen. | In conjunction with P3 and P5 the case study brochure could be extended to explain the marketing techniques used by this organization. In addition you should analyse why these techniques might have been chosen. |
| **Distinction** | | |
| D1 | Evaluate the concepts and principles applied to the marketing of products by a selected organization and make recommendations for improvement. | In conjunction with P1 and M1 you could produce a short report that evaluates the concepts and principles applied to the marketing of products by one of the organizations identified in P2, making recommendations for improvement. |
| D2 | Evaluate the marketing techniques, research and analysis used by a selected organization and make original recommendations for improvement. | In conjunction with D1, and based on the findings presented for M3, the report could be extended to evaluate the marketing techniques, research and analysis used by this organization to make original recommendations for improvement. |

**CHAPTER 3**

Courtesy of iStockphoto, weicheltfilm, Image# 6905106

Projects form the basis of a range of key business activities. Working across departments and functional areas and also consultation with third parties and stakeholders can ensure that knowledge and expertise is captured and utilised effectively into achieving a specific task or objective.

Projects can be quite small involving one or two people in the planning and implementation of a given solution to a set budget. Projects can also be very large and complex, involving a number of different people and teams, extending over months or even years and at the cost of thousands, possibly even millions of pounds.

All projects should follow a particular lifecycle to ensure that various resources are available to support each stage of development. The use of IT can indeed support all stages of a project, to facilitate human resources in delivering an achievable and satisfactory solution.

# Chapter 4

# IT Project

Working within an IT environment requires people to be involved with projects and team working. The world of IT is so dynamic, interchangeable and complex that tasks dictate a project environment, rather than individuals working in isolation. Typical IT projects may involve the development of a new website or database, network installations or upgrades, or projects that look at the feasibility of an IT proposal or system.

Working within a project environment demands a certain skill set and an awareness of a project life cycle, resource issues, influences and impacts that can decide the future success or failure of the project.

This chapter will provide an overview of the skills required to succeed within a project environment, including an awareness and development of analysis, synthesis and evaluative skills and also an ability to work independently.

This chapter will provide coverage of the following learning outcomes:

● Understand how projects are specified and managed.
● Be able to plan an IT project.
● Be able to implement an IT project.
● Be able to test, document and review an IT project.

Embedded within each section is a range of activities that will strengthen your understanding of the subject matter and provide you with the support you need to effectively complete some of the evidence requirements for the unit.

## Understand how projects are specified and managed

Numerous inputs and processes go into developing a project. The way in which a project is specified and managed can have a tremendous impact on its success. The areas that will be explored further within this section include project specification criteria, introduction to the project life cycle and project management tools, an overview of resources and other influences on a project, and finally an introduction to different project methodologies.

### Project specification

When a project is being specified and the parameters are being drawn up, a range of criteria and resources needs to be taken into consideration, as identified in Figure 4.1.

Stakeholders are people who have an interest in a particular project, organization or proposal. Stakeholders could be:

- shareholders and investors
- environmentalists and associated agencies
- patients, customers or clients
- external bodies such as the government or local authorities.

Within any given project environment there will be a range of stakeholders who are all working towards achieving specific aims and objectives.

For example, within an organization stakeholders could be people such as finance managers who control the budgets, human resources who control the staffing resources, and marketing people who oversee the promotion of the final product or service being created.

Business case requirements outline the resources and requirements of a project and can help to clarify the project's aims and objectives, problems and solutions. A business case can also help you to identify the risks involved and ways of addressing these, thus allowing for successful project implementation. If the project has to be approved by a sponsor a business case can also support you in structuring a framework within which the project can be measured and evaluated.

Setting clear objectives and having a set of clearly defined deliverables can affect the success of a project as all parties can see what has to be done, how, when and by whom.

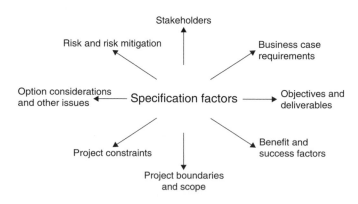

Figure 4.1 Project specification factors

Benefit and success factors can be measured against various targets that have been set within the project. If targets are being addressed and the objectives and deliverables are being met then the benefit and success factors will also improve. Success factors can be measured against time, cost or resources, for example how many people it took to implement the project.

All projects will have a boundary; however, this may not be clearly defined. The scope of the project can include being limited to a department of an organization, multiple departments, a branch or the entire organization.

Project constraints can be identified at the start of a project and should be included within any initial project proposals or specifications with recommendation on how to address these. Some constraints are unforeseeable, such as a delay in materials to complete a product or the project sponsor changing their mind on a particular aspect. As long as a workable solution or compromise can be agreed, project constraints can be managed effectively with minimal disruption to the final project design.

Options and the consideration of options are always likely to play an integral role within any project. As a project starts off decisions may be taken to change certain aspects of the team, design, time-frame, cost or other resources. An awareness of ethical and sustainable issues should also be taken into account before the project goes 'live', as this may affect where materials are resourced from, product design, packaging and distribution.

Projects should be driven by the need to succeed; however, failure should not be ignored. Having an awareness of the consequences of failure should drive members of the project team to invest more time and effort into completing the project.

With any given project there is an element of risk; therefore, risks should be identified early in the project planning stages and strategies should be formulated to try to overcome these. Some risks are unforeseeable, especially if there is a certain amount of dependency on third parties.

## Activity 4.1

### Specifying a project

Think about a task/project that you have recently completed or that you are about to start. Using the table, complete each section to identify the impacts and influences on projects and how they are specified.

| Specifying a project | |
| --- | --- |
| What people/stakeholders were or are involved in your project? | |
| What were or are your specific objectives? | |
| What were or are your project boundaries and constraints? | |
| What issues did you have/have you got to consider (ethical, sustainable, etc.)? | |
| How would you rate the risk of your project from 1 to 5? | 1 Very high risk<br>2 High risk<br>3 Medium risk<br>4 Low risk<br>5 Very low risk |

CHAPTER 4

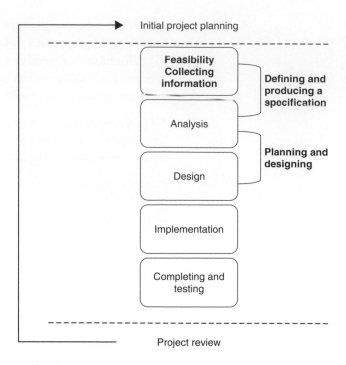

**Figure 4.2** Project life-cycle model

## Project life cycles

There are many factors to consider when taking on a project and working within a project environment. Depending on the scope and nature of the project, a number of people may be involved and the project tasks may be subdivided and allocated to teams rather than individuals.

All projects follow a life cycle that extends from the initial investigation through to the final evaluation. A traditional project life-cycle model can be seen in Figure 4.2.

One of the most important stages in any project is gathering the information, because without information you do not know what the objectives, problems and requirements are, and therefore you cannot formulate a feasible solution. In addition, without information you may be designing a project solution that falls outside the boundary of a specific price, time-frame or the use of specific resources.

To ensure the success of any given project, it should be based on a very strong framework such as the one shown in Figure 4.2. Each of the stages engages the individual or team in planning, gathering evidence and investigating, analysing, designing, implementing, and finally reviewing and testing the final project.

---

### Activity 4.2

Projects follow a natural life cycle, as shown in Figure 4.2. Think about an activity that you will need to complete in the future, for example an assignment, and set out what tasks you will undertake for each of the stages listed:

(i)  collecting information – feasibility

(ii)  **planning**
(iii) **design**
(iv)  **implementation**
 (v)  **completion and testing.**

## Project management

Successful project management can be attributed to a number of different factors and influences, one of which is the use of project management tools. Project management tools can support you in the planning and design of a project. Various tools are available, including recognized project models and a range of specialist project management software packages that can provide scheduling, planning and costing functions. These tools include Gantt charts, programme evaluation and review techniques (PERT) charts and critical path analysis. Software such as Microsoft Project can also be used to provide planning support.

### Gantt charts

A Gantt chart shows the sequence and timing of events for a particular task-based project. It identifies and predicts when activities will start and end, and examines the ordering of particular tasks (Figure 4.3).

### Activity 4.3

The Gantt chart provides a ten-week overview of how an activity can be planned and delivered.

1.  Look at the Gantt chart provided and identify whether:
    - too much or too little time been allocated to a task
    - the sequence of activities is set out in a logical way
    - any activities have been missed out (based on your own experiences of planning an activity).

    Produce a revised Gantt chart based on your analysis.
2.  For a given assessment produce a Gantt chart that illustrates the planning process you will undertake for completing the work.

| | July | | | August | | | | September | | |
|---|---|---|---|---|---|---|---|---|---|---|
| Week no. | 1 | 2 | 3 | 4 | 5 | 6 | 7 | 8 | 9 | 10 |
| **Activity** | | | | | | | | | | |
| Carry out investigation | →——————→ | | | | | | | | | |
| Collect all the evidence together | | →———→ | | | | | | | | |
| Analyse the results | | | | →———→ | | | | | | |
| Write up results | | | | →——→ | | | | | | |
| Provide some recommendations | | | | | | →———→ | | | | |
| Present findings | | | | | | | | →——→ | | |
| Gather feedback | | | | | | | | | →——→ | |
| Review the process | | | | | | | | | →——→ | |

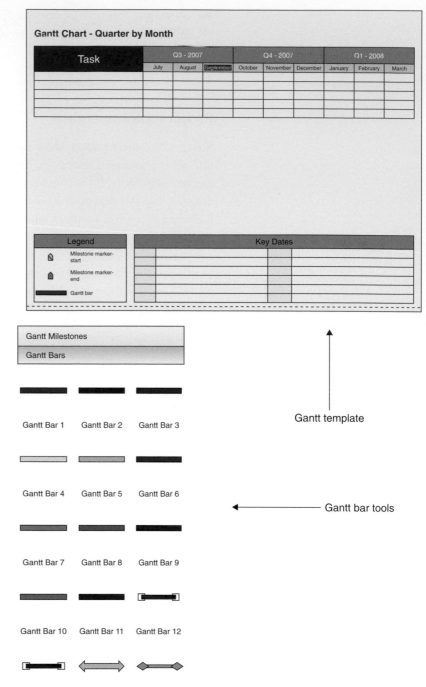

Figure 4.3  Gantt chart template created in SmartDraw

### PERT charts and critical path analysis

These two project modelling tools provide a visual overview of the scheduling, organization and coordination of tasks within a project, each providing a breakdown of tasks and their dependencies. Some tasks within a project have to be addressed in a specific order. For example, you cannot physically design a system until you have the physical hardware and software components, therefore acquiring these will impact on when the design phase takes place.

An example of a PERT chart template is shown in Figure 4.4 and a completed PERT chart in Figure 4.5.

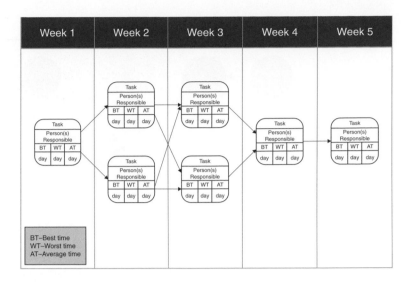

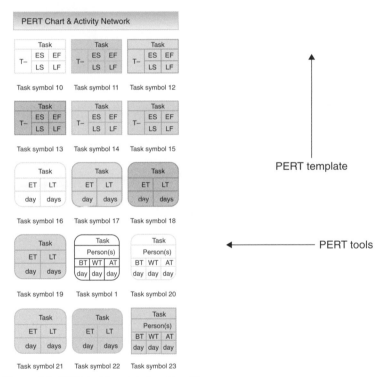

**Figure 4.4** PERT chart template from SmartDraw
http://www.smartdraw.com/

Critical path analysis works on the same principle as PERT; however, the modelling is based on a series of nodes that provide the visual scheduling function of tasks/activities.

## Resources

Projects are dependent on a wide range of resources, and the success of a project is dependent on how well these resources work together. Resources that are required to support a project include:

- information
- people

## FILM PRODUCTION

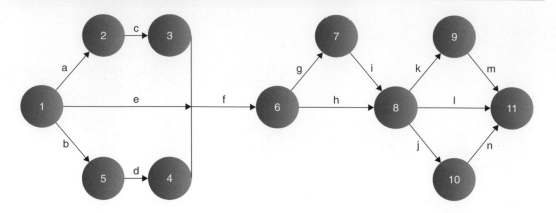

| ACTIVITIES | |
|---|---|
| CODE | MEANING |
| a | Obtain funds, loans, investors |
| b | Solicit director interest |
| c | Draw up staff contracts, agree on salaries |
| d | Pick and hire production staff |
| e | Advertise, contact agents |
| f | Scout locations |
| g | Build sets |
| h, i | Film scenes |
| j | Pick a conductor, choose songs |
| k | Edit film |
| l | Write press releases, buy ads, create preview |
| m | Prescreen with audiences |
| n | Create soundtrack CD |

| EVENTS | |
|---|---|
| CODE | MEANING |
| 1 | Obtain script |
| 2 | Budget acquired |
| 3 | Talent hired |
| 4 | Production staff hired |
| 5 | Director signed contract |
| 6 | Locations picked |
| 7 | All sets final |
| 8 | Filming completed |
| 9 | Film edited |
| 10 | Soundtrack complete |
| 11 | Film released |

Figure 4.5  Completed PERT chart

- equipment and facilities
- money.

Information is vital to any project; initially, information is required to start the planning of the project in terms of what is required, how and when, general fact-finding and early scheduling. Throughout the project information is required to ensure that targets are being met and objectives fulfilled. Finally, information is required to ensure the smooth transition of the final system, thus ensuring compatibility, reliability and the overall success of the project.

Depending on the scale of the project, one or several people could be involved, such as project managers, product developers, programmers and systems analysts. Large projects may require a range of expertise because of the complexity of tasks and skills needed, therefore people may be required who are good at organizing and planning activities, or who have technical skills such as web design, programming, hardware or networking skills. Smaller projects may only require the expertise of a single person who is able to multitask.

The success of a project is also dependent on the equipment and facilities that are being used. If, for example, you were working on a project to network an entire organization, the hardware and software used would be critical to the success of the project, with regard to whether they are priced within the proposed budget, compatible with any existing systems, user friendly, etc.

Money and budgets can usually make or break a project. Almost all projects, especially commercial projects, are driven by cost; therefore, it is crucial to monitor progress at each stage of development to ensure that finances do not constrain progress and completion.

## Other issues

Numerous factors and influences need to be considered when starting a project. Some of these have been addressed, such as information, people, equipment and money, but a range of other issues should also be considered.

Projects can be quite volatile because some aspects may be beyond your control and reliant on third parties and other external factors. Changes in external factors can have serious ramifications on a project, for example dependency on weather or climate control.

The need to monitor progress and take corrective action if activities are behind schedule or running over budget is also essential in terms of project management.

Communication is vital throughout any project; this communication could be between the members of the project team, relaying progress to a project sponsor or end-user, or liaising with third parties such as suppliers or contractors.

For some projects there may be specific guidelines or legislation that require an element or elements of the project to comply with areas such as health and safety or data protection. Guidelines on how to install equipment or hardware safely, how to transfer data between systems, and the use of protective equipment such as antistatic mats or bands may need to be considered.

CHAPTER 4

Conflict can arise within a project if people are working towards different goals or trying to satisfy different stakeholders. If people have different agendas conflict may arise in terms of prioritizing tasks or assigning capital to certain areas.

The impact of the project and its outputs can have a tremendous effect on other systems, staff and organizational structures. Introducing a new system, for example, could cause compatibility problems with other systems in the organization, software may have to be updated or revised, end-users may need to be trained and the structure of the organization may alter to reflect a new system: there may be staff redundancies, promotions or new roles created.

One way to address the impact that a project may have on a certain resource is to create an 'impact analysis' model (Figure 4.6) to determine whether the project would have a positive (+ve) or negative (−ve) influence on that particular resource. To carry out an impact analysis a number of issues should be identified, from which a table can be compiled to identify whether the impact is positive or negative (Table 4.1).

## Project methodologies

Project methodologies provide a structured framework for formalizing each stage of project planning, design, development and implementation. The advantage of using formal methodology is that you can clearly see and define what process or task has to be completed when, before proceeding to the next stage. A drawback of this is that the project can become very disjointed and prescriptive, and the wider, more holistic view of the project could be lost within all of the rigour.

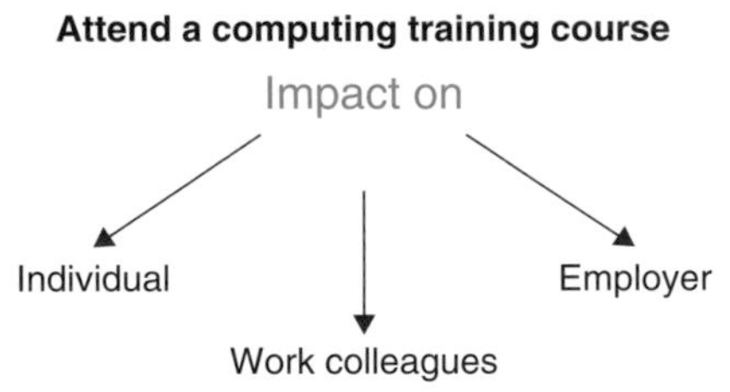

**Attend a computing training course**

Impact on

Individual          Employer

Work colleagues

Figure 4.6  Example of an impact analysis model

Table 4.1  Example of an impact analysis table

| Attend a computing training course | | |
|---|---|---|
| Area examined | Impact | Rationale |
| Individual | +ve | Learn new skills |
| | −ve | Heavier workload at work due to being out of the office to attend the training |
| Employer | +ve | Knowledge and skills can be used within the workplace |
| | −ve | Cost of the training and release from work |
| Work colleagues | +ve | Opportunity to share information and training material |
| | −ve | May feel resentment due to release from work to do the training |

Project methodologies that can be used include Prince2, Sigma and a host of company-specific frameworks.

Prince2 (Project in Enclosed Environments) is a structured method that is used for managing a range of projects on different scales, both small and large. The Prince2 framework is based on eight key processes, laying out the standards as to how they should be implemented (Table 4.2).

Sigma (Sustainable, Integrated Guidelines for Management) was established to provide support to organizations from the British Standards Institute. The guidelines provided by Sigma consist of a set of principles that help organizations to understand about sustainability and their contribution to this, and a management framework that integrates sustainability into an organization's core processes and their decision-making process.

Sigma is the first of its kind, and offers flexible and workable solutions that can be implemented into any organizational context. The Sigma Project toolkit provides further guidance on assessing opportunities and risks, reviewing performance and looking at issues facing management, and also provides support on identifying stakeholders and their level of engagement, which is very useful in the early stages of project development.

Some organizations may have developed or inherited their own project methodology. If an organization is very project driven, then a methodology may have evolved over many years, to the point where the framework is very robust and supportive of a range of organizational needs and business contexts.

## Be able to plan an IT project

Project planning consists of a number of phases that are all interrelated and heavily resource driven. A project plan provides a framework on which to plan all of the tasks and activities that will be undertaken, to ensure the future success of the project.

Table 4.2 The Prince2 framework

| 1 | Directing the project | Defining of the project by project sponsors or the senior management team |
|---|---|---|
| 2 | Planning the project | Overall planning and sequencing of tasks and activities |
| 3 | Starting the project | Following approval the resources need to be organized, a project plan devised and team members allocated |
| 4 | Initiating a project | Devising the strategy and setting criteria that will be used to evaluate the success or failure of the project at the end |
| 5 | Controlling a stage | Monitor, control and problem-solving activities to ensure that objectives are being met |
| 6 | Managing the product delivery | If different teams are working on different aspects of the project, this stage is crucial to ensure that everybody is working together and meeting the objectives and time-scales |
| 7 | Managing the stage boundaries | Ensuring that at the end of each stage any issues or problems are reported to reduce any negative impact on the next stage |
| 8 | Closing a project | Closure of the project, providing reports and feedback, checking that all objectives have been met |

CHAPTER 4

## Project plan

The purpose of a project plan is to provide detailed information about all aspects of the project in terms of tasks and activities, resources, costs and benefits, the setting of clear objectives, and a rationale for the design and implementation of the proposed system, etc., fully justifying how, where, why, what and who.

A project plan can differ between organizations and even between different contexts and environments, but there will be similarities in its content. For example, it will clearly identify phases, activities and timescales that will be measurable against a set of given objectives. In addition, all of these will be reviewed at certain time-points, and consequently action plans may arise if the review identifies that the project is behind schedule or over budget, for example.

Devising and implementing a project plan may also require the use of specialist project management software such as Microsoft Project or a scheduling package such as Prince2. The use of applications software such as spreadsheets, drawing tools, graphics and databases may also be required to ensure that the project plan is presented in a clear and professional format.

### Activity 4.4

Project plans provide an excellent framework for planning, designing and implementing a project. Project plans can be produced using specialist software packages, or they can be produced quite simply using a spreadsheet or tables.

Develop a project plan that can be used to support your project option for your module assessment.

## Detail of activities

Planning a project of any kind requires the support of a project plan that details the activities or tasks that will be undertaken.

The activities in the project plan need to be analysed in terms of parallel or sequential processing, how each process will be ordered, whether it will run simultaneously alongside another process, whether it will naturally follow a particular process, whether there are any process dependencies, etc. In addition, resources need to be identified and assigned to certain activities to ensure that there is no shortfall in requirements that could impact on the later stages of the project. Once the activities are operational and the tasks are being completed, it is essential to set review points or milestones from which you can assess the performance, target and deadline of that particular activity. By setting review points you can gain an overall picture of how the project is progressing and make predictions about the rest of the project and whether or not it will be completed to schedule. At these review stages information can also be collected that will allow you to make value judgements about the requirements for the following activities and tasks.

Information gathered may result in targets being revisited, budgets being revised and timescales being adjusted.

## Be able to implement an IT project

Implementing a project can be a complex task as it involves a number of phases, and relies on the success of these phases to ensure that the project meets its deadlines and achieves its original objectives.

### Design

To design a successful and feasible solution you need to consider the needs of end-users, as well as the requirements of the original project brief. To design a feasible solution, consideration should be given to the following questions:

- Has the problem domain been fully investigated?
- Has adequate time been allocated to each stage of the project?
- Are there adequate resources available to produce a detailed and complex solution specification?
- Have alternative solutions been considered and on what grounds have they been rejected in favour of the actual working solution?
- Is there adequate documentation to satisfy decisions made during the solution specification process?

One appropriate method that can be used to design a solution to a problem is by developing a 'requirements specification'. A requirements specification outlines the deliverables of the project in terms of:

- aims and objectives
- investigation (fact-finding) methods
- analysis of the problem domain
- proposals specification:
  - input, processing and output needs
  - design tools and techniques
- implementation and test strategies,

all of which feed into the overall design solution, in terms of its functionality, suitability and compatibility with other systems.

#### Aims and objectives

The task-related problem can be structured around aims and objectives: identifying what you want to achieve and how you are going to achieve the task(s). These can provide a framework that set out both short-term targets (for each stage of the project) and long-term targets (for the life of the project).

#### Investigation and fact-finding

Whatever the source of the problem, a thorough investigation should take place to ensure that all the necessary information has been gathered to aid the project design. There are several established investigative

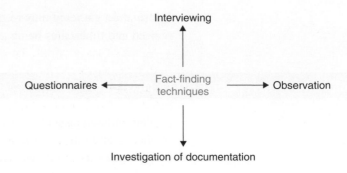

Figure 4.7  Fact-finding techniques

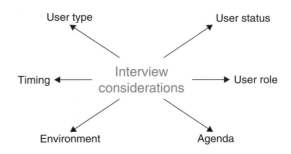

Figure 4.8  Interview considerations

techniques that aid data collection, which are sometimes referred to as fact-finding techniques (Figure 4.7).

### Interviewing

Talking to and interviewing end-users is in some cases the best way to gather information about your task-related problem. Asking questions such as:

- Can you clearly outline what the problem is?
- How many and what type of users does the problem impact on?
- Does the problem impact on other systems or tasks?
- What solutions do you have to overcome the problem?
- Are there any constraints that could impact on any proposals given?

will provide you with a detailed overview of the environment and issues relating to the task. Interviewing may also give you the opportunity to prioritize what needs to be done and when.

By interviewing users, you can ensure that the information already received is correct. Furthermore, interviewing may uncover new information and give you an opportunity to understand the system better through the eyes of the user.

A number of factors should be considered when interviewing users (Figure 4.8):

- User type – are they senior management, head of a department, team leader, data-entry clerk or administration staff, etc.?
- User status – people higher up in the hierarchy of the organization may have more limited time to assist with an interview.

- User role – what position do they have within the organization and what impact do they have on the system investigation?
- Agenda – users within the system may have their own reservations about changes to their system and may therefore be biased in their interview answers. It is up to you to decide what is fact and what is fiction. This can be achieved by validating information with another user or a third party.
- Environment – will influence the quality of information given by a user. Users will feel more comfortable in certain environments in which they feel safe and are sure that they can talk in privacy.
- Timing – if the interviewee is prepared, and has made an effort to set time aside for questioning without interruptions, the information given will be of a better quality and more detailed.

### Questionnaires

Questionnaires may provide a more convenient way of collecting information, because it can be difficult to pin down an end-user for questioning.

The benefits of using a questionnaire include:

- documentary record of wants and needs
- flexible and convenient
- can be given to a number of users simultaneously
- mass data-collection tool.

Questionnaires are an excellent way of gathering and consolidating information, providing the following conditions are met:

- The questionnaire is structured appropriately.
- A control mechanism is in place for gathering the questionnaires.
- The correct user group has been targeted.

The questionnaire should be set out clearly to provide opportunities for both short answers based on facts and figures and descriptive answers. A balance of questions will ensure that you collect all of the information required to continue with your investigation.

It is always best to provide a time limit for the return of questionnaires, for example, 'Please return within three working days'. Another way to ensure that the questionnaire is returned is to ask users to fill them in, then collect them at the end.

When designing a questionnaire another factor to consider is who the questionnaire is aimed at. The target audience is very important because different users can interpret a question very differently depending on their status and the role they play within the system (Figure 4.9).

### Observation

For some task-related problems that are quite dynamic it may be more convenient to observe the end-user in their actual role. Observation will enable you to see first hand what is going on, what issues exist, and the type of environment and conditions under which a new system would need to operate.

| ID number: | 001 | System Objective: | **Upgrade computers in the Finance and IT departments** |
|---|---|---|---|

| Name: Daniel Browne | Department: Networking | Job Title: IT Support Administrator |
|---|---|---|

Tasks undertaken each day:

- Remove backup disks and take them offsite
- Set up new users on the system
- Set up security on the file systems
- Produce system documentation and procedures manuals
- First line support – help desk
- Assist with installations and upgrades

Communicates with: Network manager, other IT support staff in the department, users at all levels, software and hardware manufacturers

Documents used:

- New user setup forms
- Internet access forms
- Backup schedules
- Support call log

| Constraints and problems: | User solutions: |
|---|---|
| 1. Too much documentation | Make the support call log automated using a database |
| 2. Users sometimes have to make multiple requests for passwords because there is no tracking system of who applied when, and sometimes the setup forms get mislaid | Better storage system and introduce a tracking system |

Please tick the following if you agree:

Problems exist with the following:

Network ☐   Operating system ☐   Other software ☐   Inexperienced users ☐

Please identify how the above have contributed to the problems with the IT system:

Any other information:

User complaints about the time it takes to attend a callout
Network keeps crashing, especially between 8:00 and 9:00 in the morning

Figure 4.9 Sample questionnaire template

### *Investigation of documentation*

Depending on your end-user or project sponsor, you may be given a range of documentation to help you to plan your project. This may be especially true if you have a real organization-based end-user.

If, for example, you were carrying out an investigation on a specific functional department within an organization, e.g. finance, to suggest ways of increasing efficiency in the payroll system, such documentation could include:

- an organizational chart
- a breakdown of personnel within the department
- job roles and descriptions of finance personnel

- sample forms used in the payroll system
- lists of procedures that are carried out on a daily basis by finance personnel.

Investigation of the documentation in this case would give you an overall picture of what is done, how it is done and by whom. You may have to use other fact-finding methods to collect more detailed information, but at least you would have a starting point to direct your investigation.

### Analysis of the problem domain

Once the fact-finding investigation has been carried out, the next stage in the project development is to analyse the information that has been collected. Depending on how much information there is, you could present the analysis in a written format, outlining your findings step by step and documenting stages of your investigation. An alternative would be to present the analysis in a visual format, illustrating the relationship between end-users, task dependencies and the overall systems environment.

The use of diagrams to illustrate any problems in the existing system can have a number of advantages over a written analysis of the problems:

- There is no need to write large amounts of notes.
- Diagrams give a clearer overview of what is happening with the entire system.
- They can easily identify key elements within the system, such as:
  - information types
  - information flows
  - storage mechanisms
  - users.
- It is easier to identify relationships and working patterns within the system.

### Proposals: solution specification

To put forward feasible design solutions a number of factors need to be taken into consideration, including:

- how the overall system is going to look and perform based on given input, processing and output needs
- what design tools and techniques are going to be applied.

Although there may be a very obvious design solution, this may not be the best design solution for the end-user. Therefore, to ensure that all proposals and solutions put forward meet the end-user's requirements the above considerations need to be addressed. To ensure that a suitable solution design has been selected to address the needs of the project brief a simple checklist can be used to focus your mind on certain key elements of the project (Table 4.3). If the answer is yes to considerations 2–7 you have to ensure that you work within these boundaries. If you have been asked to design a system using specific spreadsheet software within a two-week period and produce a written user guide, there would be no point designing a system that would not be completed in time using a database with no user guide.

CHAPTER 4

**Table 4.3** Solution specification considerations

| Considerations | Feasible Y/N |
| --- | --- |
| Has all of the appropriate information been gathered from the initial investigation that can impact on the proposal(s)? | |
| Have any constraints been imposed on the task solution? | |
| Do you have to work to a certain deadline? | |
| Has the end-user given you a solution to work with/design? | |
| Are you restricted to using certain hardware or software? | |
| Does the solution have to be designed and implemented in a certain way? | |
| Are there any other limitations or constraints imposed on the project? | |
| Do you have adequate knowledge and skills to develop the project solution? | |

If your project brief does not specify how the new system needs to be designed and implemented you have more flexibility in your choices.

### Design documentation

Design documentation to support a project can range from draft copies of work to sample screenshots or storyboards. Appropriate design documentation may include:

- a rough sketch of a graphic
- a software template for a given page of a document
- screen savers
- storyboards
- diagrams and charts
- programming code, etc.

Document design will vary depending on the nature of the project. Some projects may require high-level technical data and reports, others may require lots of graphics and screen dumps or page designs, while others may require pages of code and structured charts.

## Implementation tools

Implementing a project can be time-consuming because it results in the culmination of a number of tasks coming together, tasks that may have been set up by other members of a project team.

Depending on the type of project, implementation may be a simple 'switching-on' process or 'rolling out' of the final product, or it may require a complex set of testing, reviewing, installing, upgrading and editing features to ensure that implementation is successful. As a result of this appropriate choices have to be made, especially in terms of timing. If, for example, the project implementation requires the downtime of 200 computers, then midweek when users require the use of these computers may not be appropriate; evenings or a weekend may be more feasible.

Implementation may also be driven by a given set of hardware or software. If, for example, the upgrade of 200 computers onto a networked system is the project aim, then appropriate hardware and software will be needed to ensure that this can be done effectively and successfully.

## Deliverables

Project deliverables during the implementation stage could be:

- product based, e.g. software application, service or system
- other deliverables such as training, technical or user documentation.

Product-based deliverables could include software application, service- or system-based deliverables. These types of deliverables result in a tangible type product such as the design of a new website, implantation of a new database, introduction of a new online insurance service or installation of a new network.

Some deliverables do not have to result in a physical product. They could be more tailored to a specific user's or organization's need, such as the requirement to train employees in a particular department, or the design of new technical documents or end-user manuals to support the upgrade of new software or a new system.

## Monitoring

Monitoring and reviewing progress is crucial within any given project. Communications must be continuous with a range of stakeholders, project sponsors, management, end-users and within the project team to ensure that requirements and objectives are being met and that any issues are identified and addressed.

Monitoring can be carried out periodically at set times within the project. This could consist of a series of interim reviews at weekly team meetings. Logbooks could also be used to document progress, dates, times and any issues identified.

Following the monitoring of a project the project plan should also be revisited and updated or revised accordingly. If an issue has been identified at a review point, the plan may need to be updated to reflect the necessary action taken. Additional resources may need to be accessed as a solution to any issues identified.

Monitoring can indeed provide a mechanism for identifying what is happening at a particular point in time and provide a snapshot of progress and developments. Although this can eliminate many of the issues that arise during the life cycle of a project it cannot compensate for unforeseen events or circumstances; however, it can provide a buffer from which an action plan can be devised.

## Be able to test, document and review an IT project

Testing is required to ensure that the overall system and each of the component parts of the system work. Testing also ensures that everything is working in accordance with the requirements of the

**CHAPTER 4**

end-user. Testing can be applied to each stage of system development and is recommended especially as certain parts of the system may be dependent on others being operational.

In terms of software testing this checking at regular intervals or stages may be referred to as module testing. Module testing should be carried out to ensure that a variety of test data has been examined under a range of conditions. Following the completion of each module, overall software testing can be carried out to ensure that each component of the application design or programme is fully functional.

Testing can also be carried out to check the functionality and compatibility of the hardware.

## Completing process

The completing process will examine and check that certain tasks have been completed. These tasks will include:

- testing
- documentation
- review
- handover and sign-off
- other support.

The completion process may extend over a substantial period owing to the number of tasks that have to be addressed and signed off.

Testing could include the final checks to ensure that the completed system is fully operational. The documentation required may include test plans, user guides and support manuals, and also any review and evaluation reports that have been generated. The final review may consist of an evaluation and justification of the entire process in terms of what went well, and meeting of objectives and targets, or it may focus on a specific element of the project implementation.

The handover and sign-off could be either an informal or a formal changeover to the project sponsor or end-user. At this stage other support may also be required, such as additional training or maintenance of the system. This could be contracted out or arranged in-house.

## Functional testing

Functional testing of a product can include using test data to examine normal and extreme conditions, or may take the form of 'structured walk-throughs', all of which should be accompanied by some sort of test plan or schedule to record the test conditions and the results.

Within the implementation and testing stages consideration should be given to the following criteria (Figure 4.10):

- data capture
- validation procedures
- data organization methods and operational procedures
- output content and formats
- user interface.

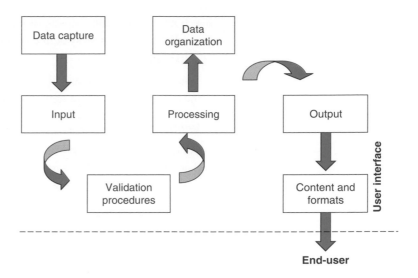

**Figure 4.10** Implementation and testing considerations

These can be addressed in terms of:

- how the data is fed into the system
- control mechanisms of the system
- how data is processed and organized
- what the system looks like.

Depending on the type of software used to develop the IT project, a number of tests can be carried out to ensure that these considerations are addressed.

### Data capture

Data capture involves the collection and inputting of data into a system so that it can be processed. Data capture can be done manually, by carrying out a feasibility study, or electronically, using a data-capture tool.

Examples of electronic data-capture tools include:

- bar codes/readers
- optical mark recognition (OMR)
- optical character recognition (OCR)
- magnetic ink character recognition (MICR)
- radiofrequency identification (RFID)
- magnetic stripe cards
- keyboard
- sensors
- CCTV.

All of these tools reflect different ways in which data and information can be input into a system. These different ways should be assessed in terms of:

- how much data is to be input and processed
- how easy to use the data-capture tool is going to be
- how appropriate the tool selected is in relation to the system specification

- how fast does the data-capture tool needs to be
- what conditions or system environment the data-capture tool will be used in.

When these factors have all been addressed a decision can be made as to the type of data-capture tool to be used.

### Validation procedures

Validation checks should be carried out during these stages to ensure that any data that is being entered and processed within the system is complete. A number of checks can be carried out (Table 4.4).

### Data organization methods and operational procedures

Data organization methods look at how the data and information within the current system is managed and what methods are used to make it into a meaningful data set. The operational procedures look at how the data set is physically managed.

During the implementation and testing stages the following areas will need to be considered:

- What existing methods and procedures are in place?
- Does the end-user require you to adapt these methods and procedures or recommend more appropriate ones?
- Is the data organized in a suitable format, e.g. does the data need filtering or sorting before designing a new system?
- Do the current operational procedures fit into the new system design or will they have to be modified; if so, is the end-user happy for you to do this?

### Output content and format

Output content and format examines what the user is able to get out of the system following processing. The output requirements may vary

Table 4.4 Validation checks

| Check | Purpose |
| --- | --- |
| Presence check | To ensure that certain fields of information have been entered, e.g. hospital number for a patient who is being admitted for surgery |
| Field/format/ picture check | To ensure that the information that has been input is in the correct format and combination (if applicable), e.g. the surgery procedure has an assigned code made up of two letters and six numeric digits, DH245639 |
| Range check | To ensure that any values entered fall within the boundaries of a certain range, e.g. the surgery code is only valid for a four-week period (1–4), therefore any number entered over 4 in this field would be rejected |
| Lookup check | To ensure that data entered is of an acceptable value, e.g. types of surgery can only be accepted from the list orthopaedic, ENT (ear, nose and throat) or minor |
| Cross-field check | To ensure that information stored in two fields matches up, e.g. if the surgeon's initials are DH on the surgery code they cannot represent the surgeon Michael Timbers, only Donald Hill |
| Check digit check | To ensure that any code number entered is valid by adding in an additional digit that has some relationship with the original code |
| Batch header checks | To ensure that records in a batch, e.g. number of surgeries carried out over a set period, match the number stored in the batch header |

between different users, for example:

- reports
- graphics – charts and graphs
- summary documents.

Consideration should also be given to how complex a breakdown is required from the output information. For example, in the case of developing a sales forecast system, the end-user may require an output screen with a month-by-month report with supporting graphs to illustrate which products performed the best and worst.

### User interface

The user interface is of vital importance to any system design. How the user will interact with the system is of prime consideration throughout the development stages. Factors that should be taken into consideration include:

- What level is the user in terms of knowledge, understanding and use of a computerized system?
- How easy is the system in terms of inputting, processing and outputting information?
- Are the screen designs clear and simple (if applicable)?
- Will the user be able to navigate around the system easily?
- Are there any integrated help/support features – menus, documentation, etc.?

If the user interface has not been designed around the needs of the user, it may well be too difficult or impractical to use.

### Test plan/schedule

The test plan provides documentary evidence that testing has been carried out and should take into account the following points:

- Version control – what test number/version are you up to?
- What has been tested?
- At what stage in the development?
- The purpose of the test.
- The results of the test.
- Comments – did the test run as expected?

Each time a test is carried out the test plan should be updated and used as a working document that can be integrated into the final evaluation. An example test plan is shown in Figure 4.11.

## Reviewing project management

Reviews should take place at every stage of a project, and a final review should be done at the end of the project to consolidate and evaluate the overall success of the project, from infancy through to maturity.

Reviews can be carried out on two levels, examining the actual project processes and stages, and the overall project management process.

A review against the specification will identify areas of success and areas that require additional development or growth in terms of the input of further resources, capital or an injection of additional time. Reviews

```
┌─────────────────────────────────────────────────────────────┐
│                    ┌─────────────────────┐                    │
│                    │  Systems Test Plan  │                    │
│                    └─────────────────────┘                    │
│                                                               │
│  Test date: ...........................  Performed by: .......│
│                                                               │
│  Stage in the systems development: ...........................│
│                                                               │
│  ............................................................ │
│                                                               │
│  Test version: (e.g. V1.0) ........................           │
│                                                               │
│  What is being tested:                                        │
│                                                               │
│  Software used:                                               │
│                                                               │
│  Test/s carried out:              Successful:  Yes      No    │
│                                                               │
│  1                                                            │
│  2                                                            │
│  3                                                            │
│  4                                                            │
│  5                                                            │
│                                                               │
│  Changes to be made:                                          │
│                                                               │
│                                                               │
│  Recommendations:                                             │
│                                                               │
│                                                               │
└─────────────────────────────────────────────────────────────┘
```

**Figure 4.11** Sample test plan

against the specification can identify developments and progress made at each incremental stage from the initial planning to the analysis, design, implementation and testing stages.

A review of the actual project management will identify the following:

- actual dates achieved set against planned/proposed dates
- actual resources used set against planned/proposed resources required
- actions taken against unexpected external factors
- the validity of tools used.

## Technical documentation

Documentation forms a critical part of any project. Documentation can include a range of technical documents that are appropriate to the project (Table 4.5) and user guides that can be classed as technical or non-technical depending on the system that has been designed.

## User guide

A user guide will help to support the end-user in the use and understanding of the system that has been developed. The guide should be set out to reflect the knowledge and level of the user, for example a non-technical document for entry-level users who have little or no experience of working with computerized systems. For a more advanced user a more technical guide could be produced outlining some of the more advanced features of the package you have used.

A user guide should include:

- instructions on how to use the product or service
- getting help

**Table 4.5** Possible documentation requirements

| Document | Use |
| --- | --- |
| Interview question sheet | To gather information on the task-related problem |
| Questionnaire | |
| User catalogue | To record the wants and needs of users in respect of the new system |
| Requirements catalogue | |
| Draft plans | To demonstrate project development and the generation of ideas and proposals |
| Designs | |
| Screen shots | |
| Implementation plans | To check that the system meets the user's requirements and works to agreed standards |
| Test logs | |
| User guide | To support end-users, providing help and advice |
| Training booklet | |

- identification of any known bugs or faults
- obtaining and using feedback
- other requirements to do with hardware or software, etc.

In all cases the use of jargon should be avoided, and pictures and screen shots should be included to make the guide as user friendly a reference as possible.

It is advised that all documentation, including draft copies of screen designs, be included with your final submission, to illustrate the rationale that you have applied at each stage of the system's design. All documentation should be clearly labelled and presented in a professional format such as a report.

## Questions and review

1. Within any given project there will be a number of stakeholders. What is a stakeholder? Provide examples of stakeholders.
2. There is a number of risks associated with any project. Identify four possible risks.
3. What is meant by a 'project life cycle' and what are the stages within the life cycle?
4. Identify two project management tools.
5. Identify three different resources that are required on a project and state why they are important.
6. Discuss the benefits and drawbacks of formal methodologies.
7. What criteria or information are required on a project plan? Produce a project plan template.
8. What is meant by the terms 'parallel' and 'sequential' processes?
9. Why is it important to draw up draft designs before implementing a project?
10. Why and at what stages should project testing take place?
11. It is important to have a range of test data, for example 'normal' and 'extreme'. Provide examples of normal and extreme data.
12. Identify a range of technical documentation that could be used on a project and state what the purpose of each is.
13. In what cases might you require a user guide?

**CHAPTER 4**

## Assessment activities

| Grading criteria | Content | Suggested activity |
|---|---|---|
| **Pass** | | |
| P1 | Explain, using examples, reasons why projects can fail. | Produce a project report that integrates a number of criteria. This initial report could provide an introduction to projects, project life cycles and planning. For P1 Section 1.0 explain, using examples, why projects can fail. Information about actual examples should be referenced accordingly in the bibliography and print-outs of the examples could be included within the appendices. |
| P2 | Describe different tools and methodologies that are available to support the project manager. | Section 2.0 of the report could describe the different tools and methodologies that are available to support the project manager. |
| P3 | Describe typical phases of a project life cycle. | In conjunction with P1, include a section within the report that provides a graphical overview of the phases of a project life cycle and describe each of the phases. |
| P4 | Identify a project, collect information as required and produce a project specification. | In conjunction with P1 a section within the report could be dedicated to the project selection and specification. |
| P5 | Develop and document a project plan. | Produce a project plan that supports your project choice for P4 and develop this throughout the project development and implementation stages. |
| P6 | Monitor the project against the project plan. | Update the project plan and monitor progress against it. |
| P7 | Design a product or service based on a project specification. | You are required to produce a physical design based on your project specification for P4. |
| P8 | Implement an IT project and create a product that meets the specification. | In conjunction with P4 and P7, demonstrate through screenshots and possibly a physical demonstration (if appropriate), that the IT project you have created meets the specification set out in P4. |
| P9 | Test and review the output of a project. | Demonstrate through the use of a test plan or other appropriate testing method that you have tested your implemented solution. The testing should also be accompanied by a written or verbal review of the output of the project. |
| P10 | Review the project management process and identify successful and unsuccessful choices made and decisions taken. | Provide a written evaluation that reviews the project management process and identifies successful and unsuccessful choices made and decisions taken. |
| P11 | Create technical and user documentation. | A range of technical and user documentation should be produced. This could incorporate the test plan and a possible user guide (if appropriate). |

*(Continued)*

| Grading criteria | Content | Suggested activity |
|---|---|---|
| **Merit** | | |
| M1 | Explain, using examples, how it is possible to minimize the chances of projects failing. | In conjunction with the report for P1, carry out research and use examples to explain how it is possible to minimize the chances of projects failing. |
| M2 | Describe critical path analysis (CPA) and explain with an example how critical paths can be identified. | Within section 2.0 of the report and in conjunction with P2, describe CPA and explain, using an example, how critical paths can be identified. |
| M3 | Independently produce a project specification that takes into account the needs of all stakeholders. | In conjunction with P4, information provided within the report could be expanded to include how stakeholders' needs have been taken into account. |
| M4 | Monitor and track the progress of a project using a project plan, adapting the plan as circumstances change. | In conjunction with P6 the project plan can be adapted to take into account a change in circumstances. |
| M5 | Meet deadlines and key review dates as identified in the project plan. | In conjunction with P6 and M4 the project plan should clearly demonstrate that key review dates have been met. |
| M6 | Undertake an interim review of the project management process and identify any emerging problems. | In conjunction with P10, part of the evaluation should also review the project management process and identify any emerging problems. |
| M7 | Demonstrate effective communications with stakeholders at all stages of the project. | In conjunction with P4, P7, P8 and M3 you should provide evidence of having effectively communicated with stakeholders at all stages of the project. A section within the report (P1) could cover stakeholder communications. |
| **Distinction** | | |
| D1 | Justify the tools and methodologies used in a project. | Produce an evaluative report that justifies the tools and methodologies they have used in their project. |
| D2 | Critically evaluate the effectiveness of a project plan to support the project. | In conjunction with P6, M4 and M5 produce a short written evaluation that examines the effectiveness of the project plan in supporting the project. |
| D3 | Identify and accurately assess impact of potential risks to a project. | In conjunction with P1, where you researched why projects fail, additional evidence could be included that looks at the potential risks to a project. |
| D4 | Evaluate the potential impact of the introduction of the product or service on wider business systems, people or processes. | In conjunction with the report for P1, provide a final section that evaluates the potential impact of the introduction of the product or service on wider business systems, people or processes. |

CHAPTER 4

E-commerce has expanded over recent years as more and more consumers have reverted to the Internet to buy products and services. Although high street shopping has not been replaced by on-line purchases, e-commerce has provided access to a global market-place that offers choice, competitive pricing and flexibility.

E-commerce has given more power to consumers by breaking down the barriers offered by high-street shopping. The ability to shop around and compare prices and services offered within the comfort of your own home or working environment. To purchase items 24/7 and to have access to a wealth of information at the touch of a button has indeed revolutionised the way in which consumers and businesses approach the buying and selling of goods and services.

# Chapter 5

# E-Commerce

E-commerce has revolutionized the way in which people and organizations buy and sell products. Trading on a local and global basis for a diverse range of products and services is possible at the touch of a button.

E-commerce has no geographical or cultural boundaries as it can be utilized through the Internet, intranet, extranet or a combination of these systems. Products can be shipped over continents and payment can be instantaneous without the need for local currencies or foreign exchange rates.

This chapter will investigate the benefits and drawbacks for society of e-commerce and the technologies involved. In addition, a range of security issues will be explored from both a user and a business perspective.

This chapter will also provide information to cover a range of learning outcomes:

- Know the effects on society of e-commerce.
- Understand the technologies involved in e-commerce.
- Understand the security issues in e-commerce and the laws and guidelines that regulate it.

## Know the effects on society of e-commerce

The impact of newer technologies on products and services may be seen as positive for an organization because of the ability to sell online, reduce costs and increase efficiency; however, the consumer may perceive these changes differently.

The growth of e-commerce has contributed to different ways of buying, selling and trading for an organization (Table 5.1).

Table 5.1 Impact of e-commerce growth

| Impact on the business | Impact on the consumer |
| --- | --- |
| More streamlined and efficient service: items can be marketed through a website, transactions can be made using e-commerce and electronic payment systems | Loss of personal one-to-one service, possibly no one to speak to, all transactions are carried out electronically |
| More variety in products and services. An organization can advertise more items online in a virtual shopping environment | Greater flexibility, e.g. wider payment options, ability to shop around for the best deals and compare prices |
| Using a website to promote goods and services means that an organization does not have to have a physical high street presence, thus saving costs | Convenience: consumers will not have to leave home to purchase items; all transactions can be carried out online |

The demands of e-commerce have grown into a requirement for organizations to provide electronic and online provisions for payments, ordering and transaction services. Those that fail to deliver often suffer in terms of a shift in customer loyalty to more dynamic organizations offering these facilities.

### Social implication

E-commerce has impacted on a range of social areas and interests. Customer perspectives have changed with the growth of online shopping, banking and entertainment options. Customer attitudes have been influenced by the way in which e-commerce provides value-added services that are easy to use with vast savings on high-street provisions and minimal delivery inconveniences. Even issues over security do not seem so major when balanced against the many benefits that e-commerce has to offer.

In terms of the impact on business and society, e-commerce has opened up new doorways and opportunities that never existed under the traditional commerce framework.

Businesses are now able to offer a more holistic package to consumers that integrates elements of marketing, using a website presence as the vehicle for promoting products and services, sales and distribution. Almost any business can now set up a website, and advertise and sell products and services to the public, and depending on the size and complexity of the provision the setup costs could be reasonably small. The increase in online businesses has opened up a much wider portal for society in terms of more competitive pricing, more choice and greater flexibility.

E-commerce has impacted on customers, business and society in a number of positive ways. Opportunities have opened up in terms of product and service choice, reduced prices and value for money. In addition, smaller businesses have new opportunities to trade, without the need for costly overheads and premises or a high-street presence to remain competitive.

The growth rate of e-commerce has had a major impact socially: the 'bricks and clicks' approach of integrating high-street presence with an online provision has proven to be very successful for a large number of retailers and service providers. One of the downsides of e-commerce, however, is that because of the competitive arena, some businesses that do have a high-street presence have suffered because they cannot compete with online pricing, choice and services.

### Activity 5.1

1. Carry out research to compare and contrast the online presence of the following:
   - Amazon
   - Tesco
   - Boots
   - Toys 'R' Us.
     Try to find at least two products that are offered by three of the organizations and check to see how competitive they are in terms of pricing and delivery options.
2. If you were going to buy one of these products, which organization would you choose and why?

## Benefits

E-commerce has provided many opportunities for both consumers and organizations, including:

- the generation of a global marketplace with no geographical boundaries
- a 24/7 marketplace
- relatively low startup and running costs
- Providing a competitive edge by having an online provision
- the ability to search for countless options across the globe from the comfort of your own environment
- access to more competitive pricing, special offers, discounts and promotions
- the ability for organizations to track and capture customer data that can then be used for target marketing approaches
- for many, e-commerce provides an additional source of income, especially with the demand in online auction sites and the promotion of e-markets and e-traders.

With the support of an excellent infrastructure, the Internet and e-commerce have grown into an interactive shopping, entertainment, financial, social and educational environment that breeds and multiplies on a daily basis. Benefits for home users include more freedom of choice and access to more competitive prices.

Almost anything can now be purchased over the Internet: books, cars, houses ... even a village (see Case Study 5.1).

## Case Study 5.1

## Village sold for £1 m on the Internet

Daily Telegraph, Saturday 28 December 2002
Tony Harnden, Washington

## Activity 5.2

As shown in Case Study 5.1, almost anything can be bought and sold over the Internet. Carry out the following tasks to identify the limits of Internet trading and the implications of buying products over the Internet.

1. Using the Internet identify five unusual products or services that can be purchased online. State how easy it is to purchase these items.
2. Do you think that buying products and services over the Internet is secure?
3. Are there any items that you feel should not be sold over the Internet?
4. Are there any items that you would not buy over the Internet, and if so what?

In terms of finance, service provisions and retail, e-commerce has changed the way in which people bank, shop, learn and entertain. Some of the apparent benefits in these areas are:

- online banking
- online ordering
- online payments
- over-the-phone transactions and services, such as:
  - mortgages
  - setting up accounts
  - payments
  - orders
- automatic stock control and processing systems
- the introduction of loyalty schemes based on information gathered from barcode products and customer data.

The growth and dependence on e-commerce provisions in the commerce and retail sectors are inevitable. However, they have changed the way in which people use certain services and their expectations of organizations within this sector. Customers now expect greater choice, flexibility and convenience.

Experts such as pension and financial advisors in specialist commerce areas are being replaced by intelligent systems, and competent operators who can navigate their way through menu systems and decision tables.

The days of paying by cash or cheque for items are slowly being replaced by debit and credit card transactions.

The process of applying for a mortgage, being assigned your own advisor, coming into a bank or building society for a personal review and two-hour session has been converted into a fifteen-minute decision process.

For some, the changes that e-commerce have created are most definitely for the better. In an age when 'time is money', convenience overrides personal service.

### Uses of the internet and e-commerce for organizations

The Internet and the use of e-commerce have had a huge impact on organizations and the way in which they trade products and services.

E-commerce has opened up markets and provided new and more profound and dynamic opportunities for organizations. Organizations can use their e-commerce provision to:

- drive marketing activities – to promote and advertise
- extend and reach out to a much wider (global) audience
- set up their own website
- promote their corporate image
- provide online facilities for ordering, payments and delivery
- check and monitor competitor markets
- research new markets and suppliers
- receive direct feedback from customers via a website
- reduce overheads.

## Drawbacks

Although each sector of society – home users and business users – has seen many benefits from e-commerce there are also a number of limitations.

Some consumers are still very cautious about disclosing their personal details online to a given provider, especially sensitive information including card details and information about themselves. Some consumers are also deterred from purchasing items online because there is no human element involved. This lack of human contact can be offputting for certain consumers, especially if they would prefer to talk to somebody about product information or delivery options, etc.

Delivery costs and times and the fact they are less negotiable online is another drawback to e-commerce. Delivery costs are sometimes dependent on a minimum spend value that could disadvantage some customers who only require a small purchase. In addition, some of the delivery times may not be convenient, there may be long waiting times if products are out of stock and the actual delivery time may not be predetermined, which means waiting in for items to arrive.

International legislation can also be a drawback, especially if some products or services are for sole distribution within a certain geographical area, for example some games console software is restricted to sales within the USA. Products and services that are deemed to be a legal purchase in one country may contravene legislation in another country.

Issues with product descriptions and advertising a product with an incomplete or no description may cause problems for some consumers. However, because of the global accessibility of the Internet a large majority of consumers may just search for that product on a different site to obtain the required information.

Security issues are always going to be a problem when data is being passed over a networked system. However, companies are vigilant in terms of having secure payment systems and encryption facilities to safeguard customer payment and account details.

For organizations the costs of setting up online facilities and maintaining a website can be a burden, however necessary to keep up with market demands and competitors.

Employees of organizations may find themselves redundant of tasks or even a job as ordering and payment systems become more e-focused. Employees may also have to be retrained to work alongside this new technology.

## E-commerce entities

There are numerous organizations providing a range of services online. These can be categorized in terms of what sort of provision they have, for example:

- e-tailers, e.g. Amazon and Ebuyer
- manufacturers, e.g. Dell

- existing retailers that may have a high-street presence, e.g. Tesco and Argos
- consumer-led entities, e.g. ebay
- entities that provide information services, e.g. the BBC and National Rail
- financial sites, e.g. Esure and HSBC
- web-only providers, e.g. Egg.com

**Activity 5.3**

1. For each of the following e-commerce entities produce a table to compare three different entities within the sector:

   - e-tailers
   - manufacturers
   - retailers
   - consumer led
   - informative
   - service provider
   - financial.

2. Your comparison should be based on the layout of the site, ease of use, ability to find products or services, information available and ability to buy products online (if appropriate).

## Understand the technologies involved in e-commerce

The Internet is a physical network that links computers globally, enabling communication links that will store and transfer vast amounts of data and information. In conjunction with the Internet, e-commerce requires certain technologies to support its provision. To ensure the success of any e-commerce provision several factors have to be considered.

Traditionally, the hardware required to gain access to the Internet included a computer and a modem with access to a telephone line (dialup). However, the need for faster, more reliable connections has led to more efficient ways of accessing the Internet, including broadband and asymmetric digital subscriber line (ADSL). The impact on e-commerce is that downloads and transactions can be accessed and implemented more easily and more quickly.

In the early days, the most common method of Internet connection was through a dialup provision using a modem to physically make the communication connection. A modem would be linked by cables to the computer and also connected to a telephone line to enable data transmission.

Issues with speed led to developments being made and to the increased availability of other transmission media. Alternative methods such as broadband and ADSL, which were reserved mainly for business use because of the costs, have became more widely available to support all elements of e-commerce, especially business-to-business (B2B) provisions. The drive to have more reliable, more accessible and faster connection tools for home users and smaller businesses has led to an explosion in broadband.

The benefits of broadband are that, although a physical connection is required, the speed of transmission and data retrieval is much faster. If a networked system has been set up, although the router is physically connected to a computer and telephone line, additional computers and laptops with wireless capabilities do not need to be physically connected by cables to gain access to Internet and e-commerce services.

Broadband is a very a high-speed connection to the Internet that is always active, i.e. on. It has a large capacity to receive and send data, currently up to 10 Mbps. Broadband services are offered by a number of providers, Pipex being just one of these (Figure 5.1).

These newer more advanced ways of connecting to the Internet have brought a host of benefits, including:

- multitasking in terms of communication methods – simultaneous telephone and Internet access
- faster connection speeds
- better reliability in terms of connecting and instant connection
- permanent connection – no need to keep dialling up.

Figure 5.1  Pipex
Pipex.com

## Hardware and software

A range of hardware and software technologies is involved in e-commerce, including web servers, browsers, server software, web authoring tools, database systems and the need for programming. Other hardware and software issues to consider include storage size, portability, download speeds, and browser and platform compatibility.

### *Web servers, browsers and software*

A web server is the hardware and software that hosts websites comprising web pages. Individuals can have their own web server; however, some companies provide this service.

A web browser enables you to view the websites hosted by the web server.

Server software provides the services for the client software to connect to, for example Internet Explorer connects to web server software.

Web authoring tools allow you to design the structure and layout of web pages, and provide editing functions to enable you to update, link and change web pages. Examples are FrontPage (Figure 5.2) and Dreamweaver.

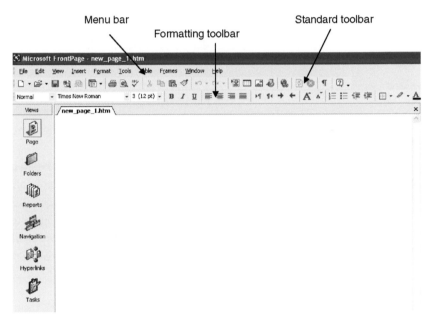

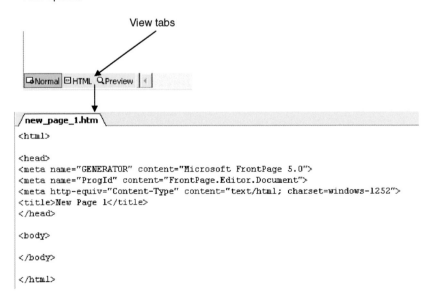

**Figure 5.2** FrontPage main screen

---

**Activity 5.4**

1.  Identify three companies that provide a web server/hosting facility.
2.  Produce a promotional leaflet that examines the type and cost of service that they provide.

---

Database systems are used by organizations and online retailers to store customer details and to track stock items and orders. Databases can also be used to track online payments and transactions. In addition, the web server software contains a database element that hosts all of the images, web pages and sounds.

Programming is required to validate data and transactions, for instance code will be required to check and process electronic orders and payments. For example, if a customer clicks on a web page button that performs an action such as 'submit details' or 'make payment', the underlying logic will be generated from a programming language/ environment such as Java.

There are several elements to consider when selecting appropriate hardware and software to support e-commerce technologies, including:

- **storage size** – ensuring that there is adequate space to store data and process information
- **portability** – ensuring that data can be transferred and communicated easily between different applications
- **download speeds** – consideration for users with slower download speeds so that they do not have to wait for your site to download
- **browser and platform compatibility** – ensuring that the site looks and performs the same across a range of browsers, operating systems and mobile devices.

## Networking

### TCP/IP

Transmission control protocol/Internet protocol (TCP/IP) is the standard protocol used for communication among different systems. It supports routing and is commonly used as an Internet working protocol. Other protocols written specifically for the TCP/IP suite include:

- file transfer protocol (FTP) for exchanging files between computers that use TCP/IP
- simple mail transfer protocol (SMTP) for e-mail
- simple network management protocol (SNMP) for network management.

The advantages of TCP/IP include:

- **expandability** – because it uses scalable cross-platform client/server architecture it can expand to accommodate future needs
- **industry standard** – being an open protocol means that it is not managed/controlled by a single company, therefore it is less prone to compatibility issues. It is the de facto protocol of the Internet

- **versatility** – it contains a set of utilities for connecting different operating systems, therefore connectivity is not dependent on the network operating system used on either computer.

### Ports and protocols

Communication between different devices requires agreement on the format of the data. The set of rules that define the format is known as a protocol. A communications protocol provides a set of rules to define data representation, error detection, signalling and authentication.

An effective communications protocol must define the following in terms of transmission:

- the rate
- synchronous or asynchronous
- full-duplex or half-duplex mode.

Protocols can be incorporated in either the hardware or the software and they are arranged in a layered format (sometimes referred to as a protocol stack) as shown in Figure 5.3. They provide some or all of the services specified by a layer in the open systems interconnection (OSI) model.

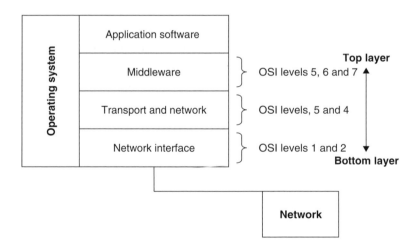

Figure 5.3  Protocol layers

### The OSI layer model

When setting up a network correctly you need to be aware of the major standards organizations and how their work can affect network communications. In 1984 the International Organization for Standardization (ISO) released the OSI reference model, which has become an international standard and serves as a guide for networking procedures and visualizing networking environments.

The model provides a description of how network hardware and software can work together in a layered framework to promote communications. It also provides a frame of how components are supposed to function, which assists with troubleshooting problems.

The OSI reference model divides network communication into seven layers. Each layer covers different network activities, equipment or protocols, as shown in Table 5.2.

**Table 5.2** Open systems interconnection (OSI) reference model

| Level | Description |
|---|---|
| 7 | Applications layer |
| 6 | Presentation layer |
| 5 | Session layer |
| 4 | Transport layer |
| 3 | Network layer |
| 2 | Data link layer |
| 1 | Physical layer |

- **Physical layer** – provides the interface between the medium and the device. The layer transmits bits and defines how the data is transmitted over the network. It also defines what control signals are used and the physical network properties such as cable size and connector.
- **Data link layer** – provides functional, procedural and error detection and correction facilities between network entities.
- **Network layer** – provides packing routing facilities across a network.
- **Transport layer** – an intermediate layer that higher layers use to communicate to the network layer.
- **Session layer** – the interface between a user and the network, this layer keeps communication flowing.
- **Presentation layer** – ensures that the same language is being spoken by computers, for example converting text to ASCII and encoding and decoding binary data.
- **Applications layer** – ensures that the programmes being accessed directly by a user can communicate, e.g. an e-mail programme.

### Domain names

It is extremely important for an organization to acquire a very memorable and appropriate domain name if it wishes to have an e-commerce or even a web presence. Most organizations will either use their company name or a shortened version as their domain name, for example:

- http://www.ba.com – British Airways
- http://www.lastminute.com – Lastminute.com
- http://www.easyjet.com – EasyJet
- http://www.train-ed.co.uk – The Training and Education Company.

Some organizations also have multiple registrations, for example .co.uk and .com

## Payment systems

A range of payment systems is available online, including:

- electronic cheque
- PayPal (Figure 5.4)

**Figure 5.4** PayPal
Paypal.co.uk

- Nochex
- credit or debit cards.

A particular e-commerce site may have a preferred method of payment for products or services, or an option from one of the payment systems listed above.

Electronic cheques are like 'floating cheques' because they allow customers to issue payments a day or so before the funds come into their account to cover the cheque. Electronic cheques are a good compromise for customers who tend not to have credit facilities.

PayPal is an easier and safer way to conduct transactions online. An account is set up and funds are linked and transferred into the account using a debit or credit card. Third parties can then have payments transferred to them without sensitive card details being exposed. PayPal acts as a buffer between the customer or user and the retailer or vendor.

Nochex (Figure 5.5) is an independent UK-based online payment company that specializes in providing support to smaller businesses.

**Figure 5.5** Nochex
Nochex.com

Credit and/or debit cards are a more direct way of making a payment online. A secure payment screen will allow the user to complete information regarding their card details and payments are authorized and taken online.

## Promotion

Promotion of any e-commerce site is extremely important and this can be achieved in a number of ways. The use of search engines is important as a large majority of end-users may turn to search engines such as Google, Yahoo or Alta Vista to conduct a general keyword, product or organization search.

Meta-tags can provide descriptions about the site that can easily be picked up by search engines. Spiders are used to feed pages through to search engines.

Promoting your site through a search engine can also mean submitting a series of keywords that describes the site or e-commerce provision so that they can be picked out by end-users as they trawl through. Some organizations also pay to have a prominent search position so that their site always appears first or highlighted, as shown in the example in Figure 5.6.

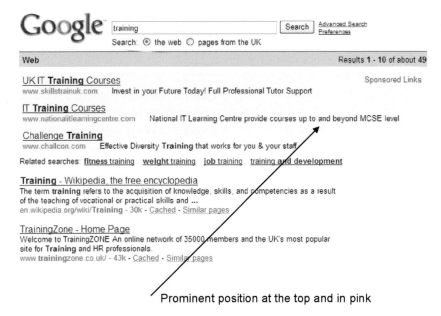

Figure 5.6 Prominent search positions
Google.co.uk

Newsgroups and forums are another way to promote a website. Some forums are extremely popular with a vast range of subscribers that check for updates each day or even many times a day. Examples include HotUK Deals (Figure 5.7) and Rpoints.

Figure 5.7 HotUK Deals
HotUKDeals.com

A range of other promotional tools and techniques can be used, such as banners and popups where miniature sites or logos appear when you log onto a website, advertising a product or service or just providing general information.

Spam is a way of reaching a mass audience by sending out images, attachments or messages advertising particular services including a site name and possibly details about who and what they are.

In terms of direct marketing it is important for an organization to have a very good, user-friendly website that can be easily navigated with an effective user interface. Regarding promotion and marketing it is important that customer loyalty is established and maintained. Some organizations do this by asking users to subscribe to a newsletter that will give them access to special offers and promotions; others use e-mail to reach their target audience by sending regular updates, etc.

## Customer interface

It is very important that any customer interface is established correctly and professionally. Any usability issues must be addressed so that users can access the e-commerce facilities easily and feel secure in their environment in terms of paying for goods and services.

Contact information should be displayed in a prominent position, in a menu or toolbar or at the bottom of the page. A number of different contact methods should be included, such as e-mail address, telephone number and possibly the physical address.

Facilities should be available to ensure that users can input their contact information easily and possibly set up an account to track the order and status, and also to store details for future reference.

A good user interface is essential so that users can access a customer services or complaints area in case they have queries or feedback regarding their shopping experience.

Frequently asked questions (FAQs) and live chat are other ways in which the customer interface can be more user orientated. FAQs can be used to address minor issues that may be easily resolved, thus saving the customer or user a telephone call or e-mail. FAQs include:

● When will my item arrive?
● What to do if I need to return an item?
● How much is the delivery cost?

Live chats are a dynamic way of discussing issues, and sharing ideas and information between two or more users.

## Understand the security issues in e-commerce and the laws and guidelines that regulate it

Various security issues should be taken into consideration when operating or setting up an e-commerce provision. As e-commerce involves trading online, buying, selling, transactions and payments, the need to conduct these activities securely is paramount as millions of pounds may be spent every minute of every day on online products and services.

## Security

E-commerce security can cover a multitude of areas (Table 5.3).

Table 5.3 E-commerce security and threats

| Security risk or threat | Details |
|---|---|
| Hacking – prevention | Ensuring that there are necessary security features, firewalls, monitoring and auditing software that will prevent individuals hacking into a computer system that they do not have permission to be in. See Case Study 5.2 |
| Viruses | A virus is a programme that has been designed to infect a system and spread into other areas or systems. A virus may corrupt, attach or delete data within that and other systems. See Figure 5.8 |
| Identity theft | Where a third party steals and uses information belonging to another person, such as their name, address and credit card details |
| Firewall impact on site performance | The primary aim of a firewall is to guard against unauthorized access to an internal network. In effect, a firewall is a gateway with a lock; the gateway only opens for information packets that pass one or more security inspections |
| SSL | SSL (secure sockets layer) provides secure communications over the Internet for services such as web browsing, instant messaging, e-mail and online transactions. One of the leading SSL providers is VeriSign (Figure 5.9) |
| HTTPS | Hypertext transfer protocol secure (HTTPS) is a secure version of HTTP: a secure way of transferring data and conducting online transactions over the web |
| RSA certificates | RSA provides an authentication/3D secure solution against fraud protection over the Internet. An example is Verified by Visa® |
| Strong passwords | On a networked system various privilege levels can be set up to restrict users' access to shared resources such as files, folders, printers and other peripheral devices. A password system can also be implemented to divide levels of entry in accordance to job role and information requirements. For example, a finance assistant may need access to personnel data when generating the monthly payroll. Data about employees, however, may be password protected by personnel in the human resources department, so special permissions may be required to gain entry to this data. Strong passwords should be used that contain a variety of alphanumeric characters to make it more difficult for people to intercept and hack into user accounts |
| Alternative authentication methods | For example biometrics and fingerprinting |

**Case Study 5.2**

**Effective ways to stop hackers**

So, what can you do to protect your tiny machine from hacker tricks? Fortunately, there are some measures that we can take, and it doesn't require us to be a Neo or Hugh Jackman's character from the movie 'Swordfish'.

These hacker protection tips are simple and effective and will defend you from most of the attacks.

### OS updates

The first thing to do in computer hacking prevention is to assure yourself that all your software is up to date; especially your operating system and your web browser. Why? Because they are the two things that hackers will try to attack first if they want to get into your computer.

### Firewalling

The second thing that you need to do is to install a firewall. As a matter of fact internet firewall hacker protection has become so necessary that Microsoft now ships it for free as part of their Windows XP operating system. It took them some years to admit it, but the truth is that their software was an easy target for the hackers and crackers that lurked through the World Wide Web.

In case you don't want to use Windows XP firewall, there are many alternatives in the market. Companies like Symantec and Zone Labs have produced software firewalls for some time and have become a necessity for all the computers of corporate America. If you don't know which one you want to buy, use the trial periods. Usually you can use the firewall for 15 to 30 days; that amount of time is more than enough to make your decision. The next step in security is to have an antivirus installed. There are free versions like AVG antivirus, or pay per year licenses, like Norton Antivirus (also from Symantec). As in the case of firewalls, there are many varieties available in the market; use the trial periods for choosing wisely.

### Anti-spyware/adware

Finally, there is the anti-spyware programme. As if viruses were not enough, companies from around the world decided to create programmes that could pick up data from your computer in order to acquire information for their databases. It may not be as dangerous as a virus, but it is an intrusion to your privacy. Wipe them out with this piece of software.

Nowadays hacker prevention has become a task for all of us. No longer is it the responsibility of the system administrator of our company. After all, he can install all the security in the world in the company's network, but if you let a virus in because of your carelessness, he won't be able to stop it. The same goes for your computer at home. You are the only one responsible for it. Remember that new hacker tricks appear as each day goes by, so you need to be prepared.

http://www.hackingalert.com/hacking-articles/hacker-protection.php

Current threats as of 3 April 2008 are shown in Figure 5.8. Onecare highlights one example of threats to systems. Another good example is McAfee, which provides a 'Threat Centre' update service.

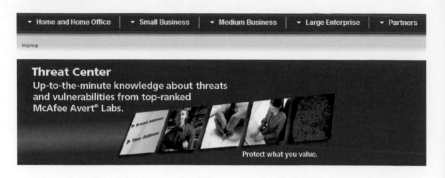

Figure 5.8  Viruses
http://onecare.live.com/standard/en-us/virusenc/

**Activity 5.5**

1. Visit the McAfee site at: http://www.mcafee.com/us/threat_center/default.asp

2. For the following identify the current top 1 and 2 of each and the date:

   - malware
   - unwanted programmes
   - vulnerabilities.

3. Look at the virus map and identify where the main geographical clusters of viruses are.

*Firewalls*

There are three basic types of firewall:

- **Application gateways** – the first gateways, sometimes referred to as proxy gateways. These are made up of hosts that run special software to act as a proxy server. Clients behind the firewall must know how to use the proxy, and be configured to do so to use Internet services. This software runs at the application layer of the ISO/OSI reference model, hence the name. Traditionally, application gateways have been the most secure, because they do not allow anything to pass by default, but need to have the programmes written and turned on to begin passing traffic.
- **Packet filtering** – a technique whereby routers have 'access control lists' turned on. By default, a router will pass all traffic sent it, and will do so without any restrictions. Access control is performed at a lower ISO/OSI layer (typically, the transport or session layer). Since packet filtering is done with routers, it is often much faster than application gateways.
- **Hybrid system** – a mixture of application gateways and packet filtering. In some of these systems, new connections must be authenticated and approved at the application layer. Once this has been done, the remainder of the connection is passed down to the session layer, where packet filters watch the connection to ensure that only packets that are part of an ongoing (already authenticated and approved) conversation are being passed.

*SSL*

The secure sockets layer (SSL) provides secure Internet connections for services such as web browsing, instant messaging, e-mail and online transactions. One of the leading SSL providers is VeriSign (Figure 5.9).

Figure 5.9 VeriSign and SSL certificates
Verisign.co.uk

## Legislation

There are many security issues surrounding the use and implementation of e-commerce. In addition, there are laws and guidelines to protect and regulate users to ensure that the e-commerce experience is much safer and less exposed.

Legislation can impact on different users in different ways. Organizations have to ensure that they operate within certain legislative boundaries, which include informing employees and third parties about how they intend to safeguard systems and any information collected, processed, copied, stored and output on these systems. Personal/home users need to be aware of the required legislation, what their responsibilities are and the penalties that can be imposed if any laws or regulations are breached.

The types of legislation of which an organization and users need to be aware include how data and information are gathered, used, processed and stored.

### Useful definitions

- **Personal data** – Information about living, identifiable individuals. Personal data does not have to be particularly sensitive information and can be as little as name and address.
- **Data users** – Those who control the contents, and use of, a collection of personal data. They can be any type of company or organization, large or small, within the public or private sector. A data user can also be a sole trader, a partnership or an individual. A data user need not necessarily own a computer.
- **Data subjects** – The individuals to whom the personal data relates.
- **Automatically processed** – Processed by computer or other technology such as documents image-processing systems.

### Role of the data protection commissioner

The Commissioner is an independent supervisory authority and has an international role as well as a national one. Primarily the Commissioner is responsible for ensuring that the Data Protection legislation is enforced.

In the UK, the Commission has a range of duties including:

- promotion of good information handling
- encouraging codes of practice for data controllers.

To carry out these duties the Commissioner maintains a public register of data controllers. Each register entry contains details about the controller such as their name and address and a description of the processing of the personal data to be carried out.

### Registering entries

All users, with a few exceptions, have to register an entry or entries giving their name, address and broad descriptions of:

- those about whom personal data is held
- the items of data held

- the purposes for which the data is used
- the sources from which the information may be disclosed, i.e. shown or passed to
- any overseas countries or territories to which the data may be transferred.

### *Trading standards*

Trading Standards (Figure 5.10) provide an information portal on consumer protection in the UK.

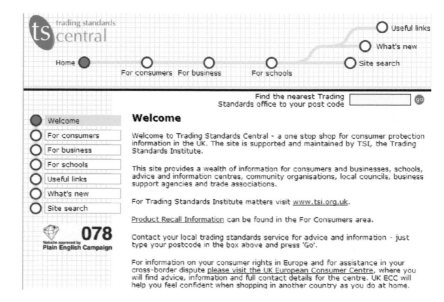

**Figure 5.10** Trading Standards
http://www.tradingstandards.gov.uk/

---

### **Activity** 5.6

1. Have a look at the Trading Standards portal: http://www.tradingstandards.gov.uk/
2. Click on the menu items for 'consumers', 'business' and 'schools' and see what information you can find out that you think might be appropriate for you to know.

---

## **Questions and review**

1. E-commerce can have a number of effects on society. Identify some of the social implications of e-commerce.
2. What is meant by the term 'bricks and clicks'?
3. Identify four benefits of e-commerce and three drawbacks.
4. What is meant by an e-commerce entity? Provide an example of an 'e-tailer', a manufacturer and a consumer-led entity.
5. What is the function of a web server, web browser and server software?
6. Provide four examples of each of the following domain names: .co.uk, .com and .org

7. Why is it important for an e-commerce site to have a secure payment system? Provide two examples of electronic payment systems.
8. Why might a user require a 'meta-tag' within their website coding?
9. Carry out research to identify two newsgroups and one forum. What do you think are the benefits and drawbacks of newsgroups and forums?
10. The customer interface is extremely important in terms of having an e-commerce provision. Identify two customer interface considerations and explain why they are important from both an organization's and a customer's perspective.
11. What measures can an organization take to prevent hacking to its site?
12. What does SSL mean? Explain what it is and what it does.
13. Identify two pieces of appropriate legislation in terms of e-commerce security.

## Assessment activities

| Grading criteria | Content | Suggested activity |
|---|---|---|
| **Pass** | | |
| P1 | Describe the social implications, benefits and drawbacks of e-commerce. | Produce a paper that will be presented at an e-commerce conference. For P1 describe the social implications, benefits and drawbacks of e-commerce. |
| P2 | Describe three different and current e-commerce entities. | In conjunction with P1, describe three different e-commerce entities. |
| P3 | Describe the hardware, software and networking technologies involved in e-commerce. | In conjunction with P1, describe the hardware, software and networking technologies involved in e-commerce. |
| P4 | Describe how e-commerce systems can be promoted and marketed. | In conjunction with P1, describe how e-commerce systems can be promoted and marketed. Include real-life examples to illustrate this. |
| P5 | Describe the security issues in e-commerce. | Produce a presentation that looks at a range of payment, security and legislation issues. It should describe the security issues in e-commerce. |
| P6 | Describe what and how legislation impacts on e-commerce systems. | In conjunction with P5, include slides that describe what and how legislation impacts on e-commerce systems. |
| **Merit** | | |
| M1 | Explain the potential risks to organizations of committing to an e-commerce system. | In conjunction with P5, presentation slides should be produced that explain the potential risks to organizations of committing to an e-commerce system. |
| M2 | Compare two different payment systems used in e-commerce systems. | In conjunction with P5, compare two different payment systems used in e-commerce systems. |
| M3 | Explain how security issues in e-commerce can be overcome. | In conjunction with P5, explain how security issues in e-commerce can be overcome. |
| **Distinction** | | |
| D1 | Evaluate the use of e-commerce in a 'brick and click' organization that balances e-commerce with a continued high street presence. | In conjunction with P1, include a section within the report that provides an evaluation of the use of e-commerce in a 'brick and click' organization that balances e-commerce with a continued high street presence. Provide real-life examples to support this – for example Boots, Tesco etc. |
| D2 | Justify the choice of security techniques used to protect an e-commerce system. | In conjunction with P5, students should justify the choice of security techniques used to protect an e-commerce system. |
| D3 | Predict and describe the potential future of e-commerce and its impact on society. | In conjunction with P1, conclude your report with a final section that describes the potential future of e-commerce and its impact on society. |

IT has influenced the way in which many businesses function and deliver their core activities. Some organisations have evolved out of the demand for IT services and provisions providing e-services, products and support.

Some organisations have had to change their business operations as a result of the impact of IT. Changes to the way in which data and information is stored and communicated across networks and issues such as security have led to

# Impact of the Use of IT on Business Systems

T he use of information technology has revolutionized the way in which businesses function on a local, national and global scale. In addition, it has impacted on work practice and communication at all levels.

This chapter will examine the way in which organizations have had to embrace and adapt to technology and how this has changed working patterns and the mode of work tasks.

Employees can now be more flexible and mobile in their roles. The emergence of networked systems and mobile technologies has enabled more people to work from home and also en route to and from work. Adapting to technology has not only created more of a financial burden to organizations that have to invest in hardware, software and expertise, but also forced some organizations to think about what they do and how they do it, and this has resulted in possible downsizing, delayering, restructuring and re-engineering employees. These issues will be explored within this chapter in addition to addressing the following learning outcomes:

- Know the IT developments that have had an impact on organizations.
- Understand why organizations need to change in response to IT developments.
- Understand how organizations adapt activities in response to IT developments.

**CHAPTER 6**

## Know the IT developments that have had an impact on organizations

A range of IT developments has had an impact on organizations. These developments have changed the business landscape in terms of how an organization functions and also can influence the future of an organization in terms of profitability, competitiveness and growth.

### Hardware and systems

The IT requirements of an organization in today's business environment dictate that hardware becomes more powerful and sophisticated, that different systems and platforms exist in a more compatible environment, and that communication methods become more dynamic, mobile and accessible to all.

Power and capacity are all important for organizations. In response to this, IT resources need to be robust enough to input, process and output data and information instantaneously and accurately at the required volumes. The storage mechanisms needed to support these system and data requirements are also required to demonstrate a high level of sufficiency.

The availability and growth of new communication technologies such as wireless, digital and mobile technologies is evident in the range of communicative IT devices now available. Wireless and mobile technologies such as laptops, personal digital assistants (PDAs) and wireless networks mean that people are not restricted to working in a conventional environment like an office. These technologies can facilitate people and enable them to work from home, on the move or in another remote location.

Networks can enhance and benefit communication systems within organizations. Information can be exchanged on an individual-to-individual need, team need or functional need internally within an organization (Figure 6.1).

In addition, networks can enhance communication outside an organization to extend to offering support to customers, ordering items from suppliers (online), and submitting financial returns and documents to third parties.

Networks can also be used for research purposes: using the Internet, as an information resource and to provide a framework for standard ways of working in terms of coordinating tasks, collaboration, informing, updating and disseminating data and information to a range of sources.

E-mail has become one of the fastest and most widely accessible formats of communication over recent years. The ability to send and receive messages has gone beyond having to sit down in front of a computer terminal and tap commands into a keyboard. New and more portable tools have become available, including games consoles, wireless applications protocol (WAP) phones, digital televisions and set-top boxes.

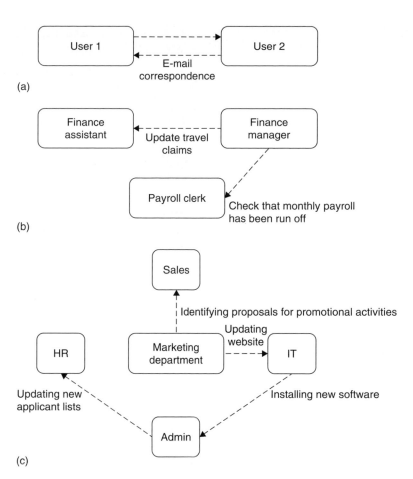

**Figure 6.1** Example network information exchange: (a) individual network communication need; (b) finance: team-based network communication need; (c) functional network communication need

The advantages of using e-mail include:

- **speed** – the ability to send messages asynchronously at the touch of the button anywhere in the world
- **multiple sends and copies** – the ability to communicate the same message to multiple users
- **cost** – minimal costs and, depending on the environment of use, possibly no cost (e.g. if sent within an educational environment)
- **convenience** – the ability to communicate twenty-four hours a day, seven days a week
- **sharing data** – because information can be transferred to multiple users, they will have access to the same shared information, which may include attached files, graphics and moving images
- **ease of use** – once familiar with some of the basic functions of an e-mail package it is very easy and user friendly (because it is icon driven) to send and receive e-mail.

There are other benefits of having and using e-mail, in that users can keep a historical record (audit) of messages that have been sent and received. Messages can be saved into different formats, updated and printed out if required. E-mail embraces a range of multimedia elements that allows users to send messages containing graphics, sound, moving images and even hyperlinks to Internet pages.

The disadvantages of having and using e-mail mainly focus on technical and security issues, such as:

- **Timing** – there is no guarantee that the recipient will read it in time to act appropriately.
- **Spamming** – the receiving of unwanted messages by advertisers, third parties and unknown users that broadcast messages universally to an entire address book.
- **Routing** – E-mail is not always sent direct from A to B; it can be routed to other destinations before it finally reaches the receiver(s). Routing can cause a number of problems:
  - more time is taken before the message reaches its destination
  - more opportunities for the message to be breached and intercepted by a third party
  - the message can become distorted or lost
- **Security** – e-mail can be intercepted easily unless some sort of encryption has been applied to the message.
- **Confidentiality** – because e-mail can get lost, intercepted or distorted it is not deemed appropriate to send confidential messages using this format.

Electronic data interchange (EDI) provides the mechanism or standard for exchanging data between two or more systems. The use of EDI has enhanced the way in which transactions between businesses, business and consumers, and business and third parties take place.

---

### Activity 6.1

Communication technologies are constantly changing to keep up with the growing demands of users.

1. Select three different types of mobile technologies and explain how these technologies have changed over the past three years.

---

## Software

As hardware has developed and become more sophisticated over the years, innovations and developments in software have also grown.

Organizational requirements now dictate more sophisticated and integrated applications that can support them in every aspect of the IT system. Specialist software that supports management, decision making and expert systems is also quite high on the list of requirements.

Software is required to support communication systems such as networks, intranets and the Internet and to keep these systems protected and secure.

Software can be categorized in a number of different ways. Operating system software is designed to support and enable the user to operate the system. Applications software provides the programmes and tools to design, model, input, manipulate and output data. This type of software includes spreadsheets, word processing, databases, graphics, presentation, communication and desk-top publishing. Finally, utility software is available to service and manage the system. This category of software is solution based, addressing issues such as security and system protection. Utility software includes printer drivers, virus checkers and partitioning software, firewalls and disk management software.

This section will give you an insight into different types of software, what they can be used for and how they can support users with specific tasks.

### Operating system software

Computers need operating system software to function. Operating systems perform basic tasks, such as recognizing input from the keyboard, sending output to the monitor or display, keeping track of files and directories on the disk, and controlling peripheral devices such as printers and scanners. In large corporate systems the operating system has an even greater responsibility in that it acts as a mediator to ensure that other programmes running simultaneously do not interfere with each other. The operating system also ensures that users do not interfere with the system, by restricting access to certain areas.

Operating systems can be classified into the following types:

- **multiuser** – allows two or more users to run programmes at the same time. In an organization some operating systems permit hundreds or even thousands of users to run programmes simultaneously
- **multiprocessing** – allows a programme to run on more than one CPU
- **multitasking** – allows more than one programme to operate at the same time
- **multithreading** – allows different parts of a single programme to run at the same time
- **real time** – responds instantly to an input command.

Operating systems such as Windows and Linux can perform a number of functional activities, as opposed to single discrete tasks. Therefore, an operating system such as Windows may embed a number of the characteristics identified within the types listed above, rather than just one.

### Applications software

Applications software enables users to interact with hardware through different text, sound, animation, video and communicative techniques.

Applications software can be defined as being general, such as programmes that can be accessed on the majority of computers, i.e. standard applications, or bespoke. Bespoke applications are those that have been designed to meet a specific end-user requirement and have been custom designed, rather than being off the shelf. Bespoke and applications software are compared in Table 6.1.

Table 6.1 Comparison of bespoke and applications software

| Bespoke software | | Applications | |
|---|---|---|---|
| Benefits | Limitations | Benefits | Limitations |
| Tailored to specific requirements and can carry out specific tasks | May take a long time to develop the software to meet an organization's needs | May be cheaper to purchase, upgrade and maintain | May not completely meet the needs of the end-user |
| Can incorporate user needs and requirements to carry out specific ICT tasks | May be more expensive to purchase because of development costs | May take less time to implement because of the uniform settings | May be of a substandard quality |
| Better quality software because it has been developed for a specific task(s) | Can be complex to use, may require retraining of the end-user | Can be easier to use because of its standard features | |
| | May not be compatible with existing ICT systems | Compatibility with existing systems | |

Applications software can be categorized into the following areas:

- word processing
- desk-top publishing
- spreadsheet
- database
- graphics
- presentation
- multimedia.

Software required to support a system can be broken down into a number of areas, including information systems and more specifically management information systems (MIS), decision support software, expert systems, Internet and intranet provisions, and security software.

Information systems are systems that have been set up to manage and support the day-to-day activities of an organization and its management. Almost every organization will have information systems ranging from a basic system relying on simple application software to process, store and deliver information required, to complex, integrated systems that support the entire organization. These include stock control and inventory, payroll and invoicing.

Information systems can be classified in terms of their function and complexity. General information systems use application software tools to process, store and deliver data and information. More specific information systems are used to support a very specialist function or need within an organization. Specific information systems can include:

- strategic-level systems
- management-level systems
- knowledge-level systems
- operational-level systems.

Each of these information system types supports various aspects of an organization from strategic and tactical levels down to operational levels (Figure 6.2). Specific information systems can also be set up and integrated into functional areas of an organization, for example there may be a sales information system or a finance information system.

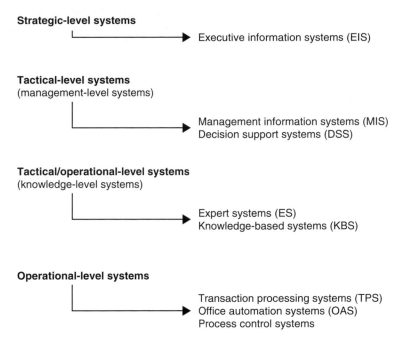

**Figure 6.2** Types of specific information system

## Strategic-level systems

This level of information system supports senior executives in making unstructured decisions at a strategic level. The types of decisions that could be made include:

- Should we consider diversifying into new markets?
- Should we make a bid to acquire new businesses?
- How could we embrace new challenges in the area of e-commerce?

Strategic-level information systems are set up to forecast, budget and plan for the future, extending over the long term, a period of five years and beyond. Within this category specific information systems can be set up, for example executive support systems (ESS).

### Executive support systems

ESS exist to support strategic personnel within an organization, their function being to provide the support and guidance needed to carry out long-term forecasting and planning. ESS use data and information collected from the current environment to establish trends or anomalies which can then be used for future planning. For example, an organization that may wish to transfer production to Europe over the next five years may look at a range of available data sources, including:

- cost of manufacturing (labour, transportation, premises)
- import and export issues (cost, initiatives, barriers to trade)
- existing businesses already trading in Europe and their profitability

- current financial status and whether there would be enough capital to finance such a venture in the future
- existing competition in Europe.

To identify specific trends, ESS may also rely on historical data to identify what has been done in the past and whether it was successful. A successful ESS will have the characteristics shown in Figure 6.3.

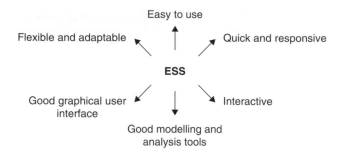

**Figure 6.3** Characteristics of executive support systems

Users who will be accessing ESS may have very limited IT knowledge or skills. Senior executives will not necessarily be technically orientated and therefore the ability to access the ESS easily and quickly is essential. Information required should be provided by the ESS within a specified period to enable further decisions to be made quickly.

An ESS must be able to interact easily and effectively with other systems to retrieve the data required. For example, decisions to be taken on whether or not to take over a new company may require the ESS to retrieve financial share price data from an external database source such as the London Stock Exchange. So that the correct decisions can be made, the modelling and analytical tools should be first class and the graphical user interface (GUI) must to be easy to use, visual and instructive.

Finally, an ESS has to be flexible and adaptable in order to continually support the ever-changing requirements of an organization.

### Management-level systems

Management-level systems are designed to support middle management in their role of making some unstructured and semi-structured decisions, but of a lower level than those offered by strategic-level systems. These systems are put in place to offer support to management levels within an organization; they are not exclusive to managers. Management-level systems provide a category in which other information systems are embedded, including:

- management information systems
- decision support systems
- operational information systems.

### Management information systems

MIS support management at all levels within an organization by providing them with data and information based on both current and historical records, from which informed and detailed decisions can be made. MIS is typically based on internal data. Examples of this are:

- financial status
- performance and productivity levels
- weekly, monthly, quarterly forecasts and trend analysis
- sales targets and figures.

The primary role of an MIS is to convert data from internal and external sources into information so that it can be communicated to all levels within an organization. Management levels will use the information produced to enable them to make more effective decisions.

### Decision support systems

Decision support systems (DSS) also support management levels within an organization, helping them to make dynamic decisions that are characterized as being semi-structured or unstructured. DSS must be inherently dynamic to support the demand for up-to-date information, enabling a fast response to the changing conditions of an organization.

DSS are complex analytical systems that are designed explicitly with a variety of analysis and modelling tools to process, enquire about and evaluate certain conditions.

### Knowledge-level systems

Knowledge-level systems are specialist systems that provide support for knowledge users within an organization. This particular category of information system is not confined to a specific user, e.g. manager, or a specific decision type, structured, semi-structured or unstructured.

The function of this type of information system is to assist an organization in its quest to:

- identify
- discover
- analyse
- integrate
- collaborate

new ideas and information, to make the organization more efficient or profitable, or to ensure high-quality standards among the workforce, services offered and/or production lines.

The users of this level of system are generally those who have achieved high academic degree or further degree status, or are members of recognized professions such as engineers, doctors, lawyers or scientists; their role within the organization being to seek out technical facts, information and knowledge, which can then be analysed, processed and integrated into the organization. Examples of how these systems can be used in a hospital include:

- identification of certain patients who are more at risk of certain medical conditions
- the impact of certain drugs on certain categories of patients
- the impact of monitoring close relatives' medical history on patients.

**CHAPTER 6**

There are many ways in which data can be extracted to provide the information required to carry out analysis or to identify the implications, impact or trend.

Some tools are quite straightforward and involve the sorting or filtering of information using conventional application software such as a database; however, there are specific tools and techniques available to serve this purpose. These knowledge tools include:

- expert systems
- data mining.

### Expert systems

Expert systems represent an advanced level of knowledge and decision support systems. Expert systems encapsulate the experience and specialized knowledge of experts in order to relay this information to a non-expert, so that they too can have access to the specialist knowledge.

Expert systems are based on a reasoning process that resembles human thought processes. The thought process is dependent on rules and reasoning, which has been extracted by experts in the field. The primary function of an expert system is to provide a knowledge base which can be accessed to provide information such as a diagnosis for a patient, to assist non-experts in their own decision-making process.

### Data mining

Data mining is a generic term that covers a range of technologies. The process of 'mining' data refers to the extraction of information through tests, analysis, rules and heuristics. Information will be sorted and processed from a data set in the hope of finding new information or data anomalies that may have remained hitherto undiscovered.

Data mining embraces a wide range of technologies, including rule induction, neural networks and data visualization, all working to provide an analyst with a more informative and better understanding of the data.

### Operational-level systems

As illustrated in Figure 6.2, this level of information system supports operational managers and supervisors, and assists them by tracking and monitoring activities that occur at this level. The categories of system that come under the operational level include:

- transaction processing systems (TPS) or data processing systems (DPS)
- office automation systems (OAS)
- process control systems.

The types of activities that may occur at this level include:

- sales figures for a set period
- production and productivity levels
- ratios examining daily work flow.

A system at this level will answer routine questions such as:

- How much is being produced on a certain basis?
- How many items are in stock?
- When will production targets be met based on current workflow levels?

Operational-level systems will provide answers to structured questions and decisions where there may be a limited number of outcomes. For example:

- How many items are in stock?

| Stock Report at 1 December 2007 | | | |
|---|---|---|---|
| Stock number | Stock item | Quantity | Location |
| RT1244000 | Fan belts | 136 | Aisle 6B |
| Y45501 | Spark plugs | 26 | Aisle 2A |
| FG2670911 | Fuses | 12 | Aisle 1D |
| HI611098 | Washers | 180 | Aisle 1B |

### Transaction processing systems/data processing systems

These systems exist to support the operational level of organizations and assist in providing answers to structured routine decisions. TPS is pivotal to any organization because it provides the backbone to day-to-day activities and processing. Examples of TPS include holiday booking systems, customer ordering systems and payroll systems.

### Data processing systems

Data processing systems carry out the essential role of gathering, collating and processing the daily transactions of an organization. These systems are also referred to as transaction processing systems (TPS). Typical functions of a TPS include:

- accounts
- invoicing
- stock management
- ledger keeping.

TPS is an essential part of an organization because it keeps its operations and day-to-day activities running smoothly and provides a base for other information support, including MIS.

Data processing systems can be characterized by their prespecified functions in that their decision rules and output formats cannot easily be changed by the end-user. These systems are directly related to the structure of an organization's data.

### Office automation systems

These systems are set up to identify and increase levels of efficiency and productivity among the workforce. To assist in this role, various tools and software are available to schedule, monitor and improve workforce activities. OAS will enable the workforce to:

- communicate more effectively
- promote collaborations and group synergy

**CHAPTER 6**

- structure daily tasks and activities
- track and schedule appointments and activities
- increase productivity by reducing repetitive workload
- automate repetitive tasks.

OAS can be quite simple, drawing on the functions of application software such as word processors, spreadsheets, databases, multimedia and communications software such as e-mail.

More complex and software tools can also be used to focus on a specific area of workflow or productivity, such as GroupWare, document imaging processing, workflow management systems or electronic document management systems.

### Process control systems

Process control systems monitor, support and control certain process activities within a manufacturing environment. Applications that are used to support process control systems can help an organization in the following ways:

- improving quality control
- assisting with project planning of the product
- assisting with physical design
- identifying resource requirements
- identifying development status or stage in the product life cycle.

A wide range of software is available to support both general and specific activities that fall under the domain of process control systems, as identified in Table 6.2.

Table 6.2 Support available for process control systems

| Software type | Function |
| --- | --- |
| Spreadsheets | Costing of manufacturing items |
| | Forecasting sales |
| | Identify break-even and profit margin points |
| | Analyse work patterns and efficiency levels |
| Statistical packages | Examine productivity levels to identify ratios of optimum working conditions |
| | Identify relationships between workforce and productivity |
| Project management | Uses Gantt charts to identify timings of activities |
| | Schedules tasks and activities |
| | Identifies task dependencies |
| Computer-aided design (CAD) | Interactive development of drawings and designs |
| | Professional drafting tool |
| Computer-aided manufacture (CAM) | Controls production equipment more accurately |
| | Integrates with other manufacturing systems |
| | Ensures quality procedures |

Internet and intranet software gives third parties connectivity to a particular website. With an intranet system the third party will consist of internal users and employees.

Security software can range from virus checkers, filtering software, encryption and moderating software to firewalls.

## Understand why organizations need to change in response to IT developments

IT developments can have a huge impact on organizations in terms of how they are structured, how they function, resources, strategy planning and their survival within the marketplace. As a result organizations have constantly to adapt and evolve as changes are implemented and in some cases imposed.

The resourcing of IT can generate many changes within an organization and present many challenges at different levels, some of which will be explored below.

### Organizational challenges

IT developments can pose a number of organizational challenges, as illustrated in Figure 6.4. Systems and processes evolve over time, and

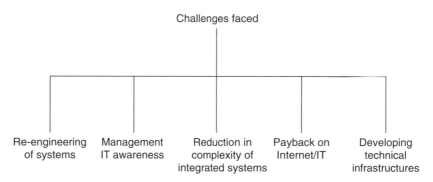

**Figure 6.4**  Organizational challenges presented by IT developments

this movement and change can be influenced by a number of factors such as the current climate of the marketplace, financial markets, growth patterns, consumer buyer patterns, supply and demand, demand for resources and IT. IT can present many challenges for organizations and as such can force organizations to re-examine and re-engineer their current systems. The need to upgrade, ensure compatibility across platforms, enhance communications, and embrace emerging and mobile technologies has re-engineered many aspects of an organization.

Management in any company should have awareness about IT and the impact that it can have on every aspect of their organization. Ignorance about IT technology can lead to a decline in profits, efficiency and productivity levels. Lack of awareness can stifle growth and prove to be an obstacle in terms of competing on the same level as other organizations within the market.

IT systems do not have to be complex and in many cases they are not; however, issues can arise when different functional areas of

**CHAPTER 6**

an organization use their own systems, some of which may not be compatible with other systems. The key to a harmonious IT infrastructure is integration and ensuring that systems are compatible, that communication is fluid throughout an organization, and that functional areas all have access to the same shared resources and information.

IT and Internet payback can be measured in a number of ways. Overall IT systems can improve efficiency, productivity and enhance communications within an organization. The Internet can increase your customer base or allow you to access new or niche markets. The company brand and image can be marketed on a global basis, and an online provision can generate a twenty-four hour market for goods and services.

Another challenge faced by organizations is to ensure that an IT infrastructure is established and that it is communicated to all users within an organization through a series of policies and training.

## External and internal environments

IT can impact on many aspects on an organization, both externally and internally.

### External environment

Externally, an organization can use IT to increase globalization and there is increased potential for competition for global companies at local levels using e-commerce.

IT and the specialist nature and technical expertise required for some elements can create potential within an organization for outsourcing and geosourcing.

IT can contribute to changes in regulatory and legal frameworks, especially with the need to ensure that data is secure and that users of IT are protected under legislation such as the Data Protection Act and the Computer Misuse Act.

IT can also reduce the costs of business startups, by providing an environment that promotes shared resources, enhanced communications, access to instant global markets and an international customer base. In addition, IT can facilitate the promotion, selling and distribution of goods and services easily and cost effectively.

---

**Activity 6.2**

Many organizations have chosen to outsource a resource or functional area as a move towards reducing costs and/or increasing the efficiency of the company.

1.  Identify two organizations that have decided to outsource an element of their provision. Explain why they have decided to outsource and whether or not this move has been successful.

### Internal environment

Developments in technology have had a tremendous impact on businesses and organizations. Organizations have had to embrace newer technologies, which has had both positive and negative implications for certain resources. With regard to employees and their work styles, developments in technology may have improved communications within an organization and generated a more interactive, collaborative environment. Technology may be more integral to a certain employee's job roles and therefore time has been saved in the carrying out of certain tasks. In addition, operations may have become more efficient and productive.

### Upskilling, training and redundancies

IT can have a detrimental impact on some aspects of an organization, especially for the employees. New technologies may require new or different skill sets, so existing employees may require training and upskilling. Some employees may be surplus to requirements if they cannot adapt to newer systems, and therefore redundancies may be enforced to make provision for new employees with the right IT skill sets.

### Home and remote working

Home and remote working has grown owing to the accessibility and falling costs of having ICT provisions in the home and access to mobile technologies. For some, the ability to work at home or on the move (while travelling to and from work) has meant that they have adopted a more flexible, possibly more productive working style. This movement towards working from home and remote working has been supported greatly by the flexibility and portability of ICT provisions. In addition, more flexible working practices have been influenced by:

- improved and enhanced communication systems such as the internet, e-mail, teleconferencing and mobile technologies allowing people to send and receive information outside the workplace
- falling costs of hardware such as personal computers and laptops, which has given rise to a new 'home office' culture
- compatibility and uniformity of some operating systems and software, thus enabling employees to work from home or remotely using the same formats and systems as they would use in the office
- initiatives by employers to supply workers with the necessary hardware and software to enable and encourage greater flexibility in working hours.

---

**Case Study 6.1**

**Out of the office, into the unknown**

Working from home sounds like an ideal solution to the problem of balancing quality of life with a career. A growing number of people are trying it and next year all employees will be given the 'right to ask'. …

… A month ago Richard Evans swapped the life of a home-worker for the hassle of a 50-minute daily commute to Nottingham. He had been working from home since

February after spending years travelling extensively for his work. While he welcomed the flexibility and lack of commuting he missed being in an office.

'You miss out on the social side of it, the coffee machine chats are an important part of working life' he says. 'There is a buzz in the office that isn't at home' he adds.

More than one in 17 British workers are now estimated to work from home. Next April the government as part of its employment bill will introduce a 'right to ask'. Nominally for parents of small children, but expected in practice by some personnel professionals to be applied across the board. It will mean that an employer will need to give serious consideration to a request for flexible working and explain any denial in writing.

The Guardian

Saturday 7 December 2002

---

## Activity 6.3

**Working from home can generate a number of benefits based on personal circumstances. The peace and tranquillity of working within a home office with minimal distractions and reduced pressure to satisfy the demands of everyday work pressures can appeal to some. However, some find that working within such an isolated environment can be a disadvantage, with nobody to interact with and exchange gossip and information with.**

1. **Identify a range of professions where home working would be a viable option.**
2. **What do you think are the personal benefits of working from home?**
3. **Would you choose to work from home in the future? Why?**
4. **Do you see any disadvantages to working from home? Are there any particular groups of people that would be more vulnerable/distracted by working at home?**

---

### Restructuring, delayering and managing change

IT can impact on every aspect of an organization and as a result changes may be required to embrace newer, more efficient ways of doing things. An example of change can be seen within the organizational fabric or structure. The structure of an organization can change: as systems become more integrated, functional areas could merge. The need for a hierarchical structure may diminish as management layers are reduced, and a process of delayering determines what and who is actually required.

Managing this process of change to the infrastructure, resources and work patterns can be a delicate task, and requires cooperation and acceptance by employees at all levels. Resistance to change can result in inefficiencies, profit losses and an inability to compete in certain markets or on a global basis.

### Employees, contractors and outsourcing

Organizations need to change in response to IT developments and one of these changes may relate to human resources and the need to access certain technical skills. Retaining a balance of core employees is essential to ensure that functional activities are carried out efficiently. Organizations that do not have the technical expertise may opt for other alternatives to expand their employee core. This could include outsourcing to other companies, consultants or individuals, or using contractors to fulfil tasks that cannot be accomplished by core internal employees.

## Understand how organizations adapt activities in response to IT developments

As IT develops and becomes more advanced, organizations have to adapt to keep up with current technology. Developments and innovations in IT should be acknowledged by organizations and in some cases embraced and integrated into their current systems.

### Activities

Organizations need to adapt to developments in IT as it can impact on day-to-day and long-term activities and strategies. Activities that may be affected include:

- sales and marketing
- purchasing
- customer support/servicing
- finance and e-commerce
- supply
- logistics
- business integration
- Internet presence
- manufacturing
- intermediation.

Sales and marketing strategies may need to adapt to developments in IT. IT can change the way in which goods and services are promoted and also change the dynamics of the customer and market base, especially in conjunction with the Internet and e-commerce. Opportunities to buy and sell goods and services on a global basis can streamline sales and marketing activities and shift the focus more towards supporting an electronic provision.

New purchasing opportunities may arise from developments in IT. Automated procedures and the use of electronic data interchange (EDI) may be used in addition to or as a replacement of existing systems.

Using new technology in the area of customer support/services can have a major impact on increasing response times, personalized messaging and providing instant access to customer data, improving the quality and overall efficiency of any service provision.

IT can also impact on other elements of an organization and change the way in which systems function and are managed. Finance and e-commerce, with secure fund transfers making transactions easier and safer; supply change management and logistics; business integration with more enhanced communications and shared resources, and having an Internet presence linked to sales and marketing strategies are all examples of activities that may be affected. Other activities to consider are manufacturing, the use and integration of automated procedures, and finally intermediation.

## Performance

Organizations have to adapt their activities in response to IT developments. In terms of performance, this can impact on:

- productivity gains
- cost reduction
- improved MIS
- better control
- improved customer service
- better synergy
- integration of systems.

With regard to performance, an organization can increase efficiency and make savings if IT is updated or introduced into various elements or activities. Systems can become more productive as tasks are automated or more easily accessed.

Costs can be reduced as overheads may be lowered or as systems become more streamlined. The quality of data and information being captured, input, processed and output may improve, thus improving any MIS in place.

Systems can be controlled and monitored more effectively through automated procedures and appropriate software. With improved systems and faster access to information the customer service provision may also improve and synergy throughout the organization may rise.

Finally, with the introduction of IT, systems will become more integrated as resources are used across networked systems, and access to data and information will become easier and more flexible.

## Managing risk

Risks to IT systems that organizations have to identify and manage include:

- cybercrime (e.g. diverting financial assets)
- sabotage to communications
- stealing of intellectual property rights
- denial of service attacks (maliciously bringing down a website)
- e-commerce threats to payments and site security.

**Case Study 6.2**

**Cybercrime: 'I Love You' virus**

The 'I Love You' virus struck more than 45 million computers worldwide in May 2000. It was one of the most widespread and damaging viruses ever. This case study will elucidate the deleterious impacts of this virus.

## What exactly did this virus do?

The virus typically arrives as an attachment in an email with the subject 'I love you'. The email, in essence, contained a message similar to this (the actual message varied):

'Kindly check the attached LOVELETTER from me.'

When the user opens the attached file (LOVE-LETTER-FOR-YOU.TXT.VBS), the virus infects the victim's computer and immediately replicates and emails itself to all addresses found within a victim's email address book.

The I Love You virus did the following to an infected computer:

- The virus deleted or altered graphic files and music files and rendered them useless.
- Redirected the victim's Internet browser to a malicious website, where another malicious programme called 'WIN-BUGSFIX.EXE' is downloaded on to the victim's computer. This is a Trojan horse, which combed the victim's computer to find passwords and sent them to an Internet account in the Philippines.

## Damage caused

Calculating damages caused by viruses is a sticky business, as it is impossible to track the exact number of people who are affected by a particular virus. Therefore, at best, only rough estimates can be made about the extent of damage of the I Love You virus.

As such, it is believed that the I Love You virus caused around US$ 10 billion worth of damages. It is estimated that the virus affected about 45 million computers world wide.

## Lesson to learn

Perhaps, one of the main reasons why the I Love You virus caused such unprecedented damage was because of its email subject. When victims saw the word 'I love you', curiosity got the better of them and they paid the price for carelessly downloading the attachment.

One important thing that email users should never do is to download attachments from strangers. And even if you receive an email from your friend, do not open it if you suspect that something is wrong about the email. Safety is better than cure, and with some cautiousness, one can avoid the hassles caused by a virus attack.

http://library.thinkquest.org/04oct/00460/ILoveYou.html

CHAPTER 6

### Activity 6.4

Case Study 6.2 highlights just one example of cybercrime that had a major impact on users across the world.

1. Carry out research to find three other examples of cybercrime.

There is a range of other risks associated with internal incompetence, hackers, viruses, Trojans, worms and natural disasters. All of these risks can threaten the physical, personnel, hardware, communication and software security of the system.

Risks that organizations need to be aware of can be categorized as:

- **physical threat**s – theft of hardware, software or data
- **personnel threat**s – staff members deleting or overwriting data
- **hardware threat**s – system crashes or processor meltdown
- **communication threat**s – hackers or espionage
- **virus or trojan threat**s – infection and propagation
- **natural threat**s – disasters such as fire, flood, earthquake or lightning
- **electrical surge or power loss threats** – overloading the system or rendering the system disabled
- **erroneous data threats** – inaccurate data in the system.

Once potential risks have been identified, policies can be put in place to address them. A corporate security policy can be drawn up to address such risks. Typical content may include:

- **access controls** – identifying and authenticating users within the system, setting up passwords, building in detection tools, encrypting sensitive data
- **administrative controls** – setting up procedures with personnel in case of a breach; disciplinary actions, defining standards and screening of personnel at the time of hiring
- **operations controls** – backup procedures and controlling access through smart cards, log-in and log-out procedures and other control tools
- **personnel controls** – creating a general awareness among employees, providing training and education
- **physical controls** – securing and locking hardware, having another backup facility offsite.

### Risk analysis

Risk analysis examines how open an organization is to security breaches based on their current security provisions. In the case of Visa (Case Study 6.3), the fact that they have been breached twice in a year will force them to examine the effectiveness of their internal and external security provisions.

### Case Study 6.3

### Fraudsters hit Visa for second time

The credit card details of 'a large number' of Visa customers in America and Europe have been stolen from a US-based retailer, Visa said yesterday.

It is the second time this year that the credit card giant has fallen victim to an attempt to illegally obtain card numbers. Last February, a computer hacker gained access to 5 m Visa and Mastercard accounts in the US.

Visa yesterday said it was cooperating with the American authorities on the matter. It also said it had issued a fraud alert to its member banks at the American retailer's database.

Although Visa declined to comment on the exact number of cards compromised because of the investigation, a spokesman for Visa Europe said: 'Everyone who used a credit card at this US merchant could have been affected.'

Danielle Rossingh, The Telegraph

11 June 2003

## Activity 6.5

**Hackers are a major risk to organizations as they become more and more sophisticated in the tools and techniques that they use to gain unauthorized access into corporate systems.**

1.  **What steps can an organization take to stop hackers gaining unauthorized access into their systems?**
2.  **Carry out research to identify what penalties are enforceable against hackers.**
3.  **Carry out research to find out what legislation has been set up to protect organizations against hackers.**

Risk analysis can identify the elements of an information system, assess the value of each element to the business, identify any threats on that element and assess the likelihood of that threat occurring. A number of preventive measures can be taken to reduce the level of risk or eliminate the risk completely; these measures can ensure both data and system security and protection.

Keeping data and systems secure can be difficult because of the environment in which users work and levels of user and access requirements to the data.

With the movement towards a totally networked environment promoting a culture of 'sharing', the issue of data security is even more important and should be addressed at a number of levels (Figure 6.5). As illustrated,

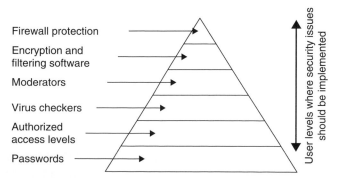

**Figure 6.5** Levels of security

security measures need to be integrated at each user level within an organization. The indication of security measure does not confine it to a certain level but reflects, on an organizational scale, what should be implemented and the scale of implementation. In addition to these proposed security measures, the option of physical security also exists – ensuring that hardware and software are kept physically secure under lock and key.

The actual protection of data can be resolved quite easily by introducing good practice measures such as backing up all data to a secondary storage device, limiting file access, and imposing restrictions to read only, execute only or read/write. Data protection is also covered more widely under certain Acts such as the Data Protection Act 1984.

### Firewall protection

The primary aim of a firewall is to guard against unauthorized access to an internal network. In effect, a firewall is a gateway with a lock; the gateway only opens for information packets that pass one or more security inspections.

There are three basic types of firewall:

- **Application gateways** – the first gateways, sometimes referred to as proxy gateways. These are made up of hosts that run special software to act as a proxy server. Clients behind the firewall must know how to use the proxy, and be configured to do so to use Internet services. This software runs at the application layer of the ISO/OSI reference model, hence the name. Traditionally, application gateways have been the most secure, because they do not allow anything to pass by default, but need to have the programmes written and turned on to begin passing traffic.
- **Packet filtering** – a technique whereby routers have 'access control lists' turned on. By default, a router will pass all traffic sent it, and will do so without any restrictions. Access control is performed at a lower ISO/OSI layer (typically, the transport or session layer). Since packet filtering is done with routers, it is often much faster than application gateways.
- **Hybrid system** – a mixture of application gateways and packet filtering. In some of these systems, new connections must be authenticated and approved at the application layer. Once this has been done, the remainder of the connection is passed down to the session layer, where packet filters watch the connection to ensure that only packets that are part of an ongoing (already authenticated and approved) conversation are being passed.

### Encryption and filtering software

Encryption software scrambles message transmissions. When a message is encrypted a secret numerical code, the 'encryption key', is applied, and the message can be transmitted or stored in indecipherable characters. The message can only be read after it has been reconstructed through the use of a 'matching key'.

### Moderators

Moderators have responsibility for controlling, filtering and restricting information that is shared across a network.

### Virus checkers

These programmes are designed to search for viruses, notify users of their existence and remove them from infected files or disks.

### Authorized access levels and passwords

On a networked system various privilege levels can be set up to restrict users' access to shared resources such as files, folders, printers and other peripheral devices. A password system can also be implemented to divide levels of entry in accordance with job role and information requirements.

For example, a finance assistant may need access to personnel data when generating the monthly payroll. Data about employees, however, may be password protected by personnel in the human resources department, so special permission may be required to gain entry to this data.

Audit control software allows an organization to monitor and record what they have on their network at a point in time and provide them with an opportunity to check that what they have on their system has been authorized and is legal.

Over a period of time a number of factors could impact on how much software an organization acquires without their knowledge. These can include:

- illegal copying of software by employees
- downloading of software by employees
- installation of software by employees
- exceeded licence use of software.

These interventions by employees may occur with little or no consideration for the organization and its responsibility to ensure that software is not misused or abused.

### User rights and file permissions

Within certain IT systems, users are given permission to access some areas of a folder, application or document and are restricted from others. By allowing users certain rights within a given system, security of data can be reassured and the span of control can be limited. An example of this can be seen in the case of an IT system in a doctor's surgery. The administration staff may have access to appointments and scheduling, the nurses may have access to patient information, and the GP could have full access and rights to print out prescriptions and authorize medication.

Certain permissions may also be set up to allow certain users partial access to a file, so for example information can be read (read only) but not written to, or users may be able to run a programme (execute only) but not view it. Users with full read/write permissions would be able to view, update, amend and delete accordingly.

### Corporate information systems security policy

Contingency plans can be used to combat potential and actual risks to a system. The majority of organizations will have an adopted security policy of which employees would be aware, the plan and policy being open to continuous review and updating. The structure of a contingency plan is unique to an organization and its requirements.

Some organizations are more at risk than others, depending on:

- size
- location
- proximity to known natural disasters and threats – flood areas, etc.
- core business activity.

A strategy based on recovery recognizes that no system is infallible.

### Disaster recovery plans

A disaster recovery plan will include provisions for backing up facilities in the event of a disaster. These provisions may include:

- subscribing to a disaster recovery service
- an arrangement with another company that runs a compatible computer system
- a secondary backup site that is distanced geographically from the original
- use of multiple servers
- use of mirroring techniques for backup and utilizing hardware redundancy to have multiple systems in place in case of failure.

Some large organizations may have a 'backup site' so that data processing can be switched to a secondary site immediately in the case of an emergency. Smaller organizations may use other measures such as redundant array of independent disks (RAID) or data warehousing facilities.

## Questions and review

1. What security measures can be enforced within an organization?
2. What is meant by the term 'risk analysis'?
3. Why are some organizations more at risk than others in terms of potential threats to their systems?
4. What measures can a large and a small company take to protect their data?

# Assessment activities

| Grading criteria | Content | Suggested activity |
|---|---|---|
| **Pass** | | |
| P1 | Describe new hardware and software technologies that impact on organizations. | Produce a comprehensive information leaflet aimed at middle management within a large networked organization. For P1 there should be a section within the leaflet that describes new hardware and software technologies that impact on organizations. |
| P2 | Identify the challenges posed to organizations because of IT developments. | In conjunction with P1, students should identify the challenges posed to organizations because of IT developments. |
| P3 | Describe how the internal environment of an organization has changed due to changes in the external environment brought about by IT developments. | Produce a report that describes how the internal environment of an organization has changed due to changes in the external environment brought about by IT developments. |
| P4 | Identify the changes in activity and performance that can arise from adapting organizational activities. | In conjunction with P3, include a section within the report that identifies the changes in activity and performance that can arise from adapting organizational activities. |
| P5 | Describe ways that organizations can manage their risks when using new IT technology. | In conjunction with P1, describe the ways that organizations can manage their risks when using new IT technology. |
| **Merit** | | |
| M1 | Explain the impact of IT developments for an organization. | In conjunction with P1, explain the impact of IT developments for an organization. |
| M2 | Explain why organizations modify their activities in response to IT developments. | In conjunction with P1 and M1, explain why organizations modify their activities in response to IT developments. |
| M3 | Explain how employees and employers are affected by changes in IT. | In conjunction with P3, include a section within the report that explains how employees and employers are affected by changes in IT. |
| **Distinction** | | |
| D1 | Make reasoned recommendations about how a particular organization can take advantage of IT developments. | In conjunction with P1, carry out research into a given/selected organization and make reasoned recommendations about how they take advantage of IT developments. This section of the information leaflet could be presented as a case study feature. |
| D2 | Assess the possible consequences for an organization in implementing IT-based changes. | In conjunction with P3, as part of the report conclusion, assess the possible consequences for an organization in implementing IT-based changes. |

**CHAPTER 6**

# Index